52 PROJECTS USING IC741

D1825120

by

RUDI & UWE REDMER

BERNARD BABANI (publishing) LTD
THE GRAMPIANS
SHEPHERDS BUSH ROAD
LONDON W6 7NF
ENGLAND

This book is translated from Topp Book No.47 entitled IC741, 1 Baustein = 52 Schaltungen, published by Verlag und Druckerei M. Frech, 7 Stuttgart 1 (Botnong), W. Germany.
©1973

English Language Rights held by Babani Press, London, England.
©1975

The Authors would like to thank, Wireless World, Elektronik, Elector and Fairchild for assistance given with certain circuits.

I.S.B.N. 0 85934 027 9

First Published – September 1975
Reprinted – July 1977
Reprinted – December 1978

Made and printed in Great Britain by
C. Nicholls & Company Ltd. Manchester

CONTENTS

EQUIVALENT SEMICONDUCTORS TO THOSE USED IN
CIRCUITS IN THIS BOOK

AD161	2N1218-1292-1722-4077,AD167,RS276-2006
AD162	AD143-152,ADY27-28,2N1537,2SB367,RS276-2006
BC107	BC182-237-582-171-547-207-129-147,RS276-2031
BC141	BC211A-300,BSX46,BSW39,2N3019-3107,RS276-2014
BC160	BC304-313-460,BSV15,2N4032-5855,RS276-2021
BC177	BC212-307-512-204-557-261,RS276-2023/2024
BC340	BC140,RS276-2014,2N3019/20-3109/10
BC340/16	BC140/16,RS276-2014,2N3019/20-3109/10
BC360	BC160,RS276-2021,2N4032-5855
BC360/16	BC160/16,RS276-2021,2N4032-5855
BD167	BC439
D13TI	2N6027,BSV58/A/B
MJ900	BDX62,2N3789
MJ1000	BDX63,2N3713
MJ3001	BDX65A,TIP141
MPF102	SK3116,2N3819-5486
2N3055	BD130,BDX10,BDY20-39,2N3713,RS276-2020
2N4443	–
2N4871	RS276-2029

uA725	LM308,RS276-1726,LM108

LDR03	RPY25,ORP12,RS276-116
LDR07	RS276-116
OCP70	OC70 with case scraped to let light enter = RS276-2004

1N4001	RS276-1101/1135
1N4007	RS276-1114,BA-133-159,BAY91,BY103-138-142

INTRODUCTION

This book has been produced in order to show the user of linear integrated circuits how easily well designed circuits can be constructed with higher reliability by application of these components. As prices have been reduced during recent years, the amateur is now in the position to use the IC 741 for his own requirements. Compared with the costs of discrete circuits, the utilisation of this component will in part be less expensive.

The inexperienced user does not realize that many different uses may be obtained with one and the same amplifier. Moreover, he may get confused because of the various type numbers different manufacturers give to the same integrated circuit. It should be mentioned that amplifiers with the same basic characteristics but higher gain and input resistance are available under different type numbers.

The author of this book deliberately shows the various applications possible with this type amplifier. After having built some circuits with the help of the following instructions, the reader will be in a position to develop his own ideas in a short time.

For the circuits shown in this book, monolithic integrated circuits have been applied in form of differential amplifiers. They are also called operational amplifiers because originally they were designed to accomplish arithmetic operations in analogue computers. Integrated circuit means that all elements of the circuit are enclosed in only one case. As the different components are eched and diffused in a chip of single critallonic material, the circuits are named monolithic amplifiers. Linearity means that the amplifier produces an output signal with the ratio between input and output constant (gain).

FUNCTION OF THE OPERATIONAL AMPLIFIER

The component has two input pins. One of them is called inverted
input and marked by minus. The other input is called non-inverted and
is marked by plus. Positive voltage if applied to the inverted or –
input, causes the output to go negative, that means the amplifier
inverts the signal. The positive voltage applied to the non-inverted or
plus input, causes the output to go positive.

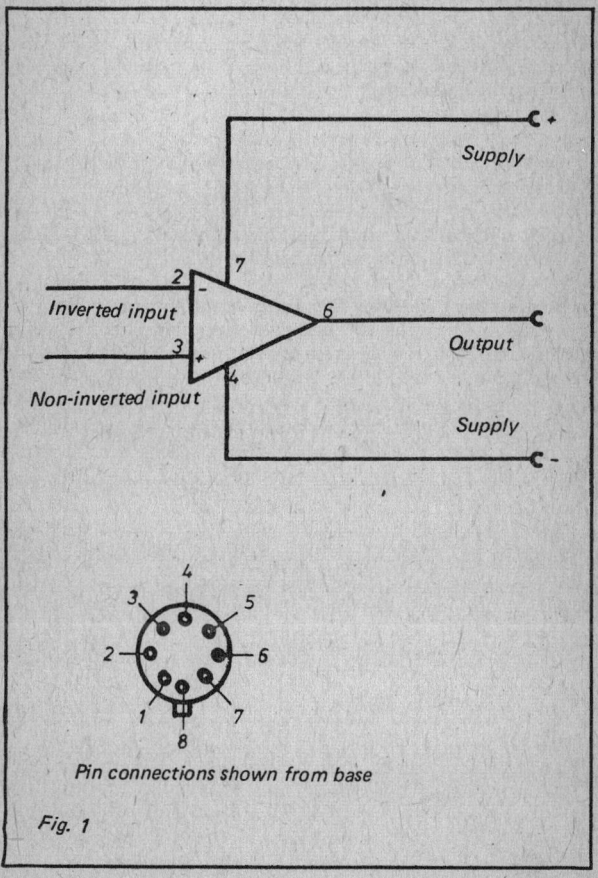

Pin connections shown from base

Fig. 1

Many readers will miss the schematic of the internal circuit, but it is nearly impossible for even an experienced amateur to understand the function of the integrated circuit, performed by many stages like differential amplifiers, constant current generators and transistors sometimes used as diodes or resistors.

It is more important to know how to use and operate this component.

Its special advantages are:

High temperature stability, high linearity of gain and small physical dimensions.

The circuit (Figure 2) and the additional truth table will explain its operational characteristics.

First open all switches. Now zero potential is applied to both inputs by the resistors A_1 and A_2. The output will show no voltage difference i.e. zero.

When the switches are operated, the related output voltage for any input condition can be found in the table.

The last two lines show that the output goes to zero even when both inputs are applied to the same negative or positive potential. Therefore only different input voltages cause the output to be positive or negative, equal voltages do not.

Truth Table				
S_1	S_2	S_3	S_4	Output
open	open	open	open	±0
+	open	open	open	−
open	+	open	open	+
open	open	−	open	−
open	open	open	−	+
+	+	open	open	±0
open	open	−	−	±0

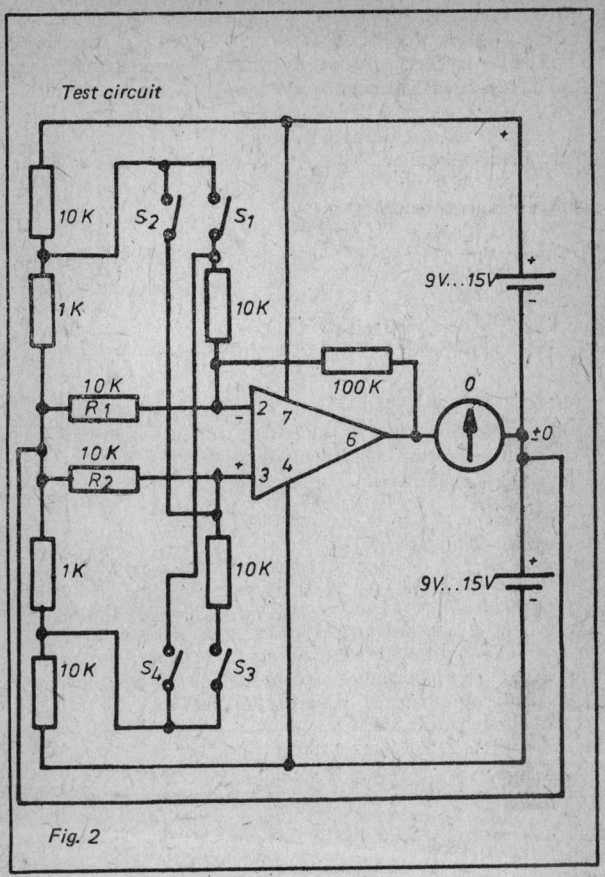

Fig. 2

DETERMINATION OF GAIN

The amplifiers described in this book have voltage gains of between 20,000 and 200,000 depending on type.

By means of reverse feedback from pin 2 to pin 6 the gain can be determined as required.

The gain is given by: $V = 1 + \dfrac{R_3}{R_1}$

e.g. $R_1 = 10K\Omega$
$R_3 = 1M\Omega$.

The gain will be: $1 + \dfrac{1 \times 10^6}{1 \times 10^4} = 101$ times .

This kind of reverse feedback is adequate for circuits with low input resistance and low gain. For amplifiers with high input resistance and high gain the resistor values would increase to Giga or Teraohms. Therefore the configuration of Figure 4 is used.

Now the gain is given by: $V = \left(1 + \dfrac{R_4}{R_5}\right) \dfrac{R_3}{R_1}$.

Of course the feedback and consequently the gain too is adjustable. In this case R_3 is a potentiometer. Often a frequency dependent gain is required, for magnetic microphones for instance, disc pick up, and so on. RC networks in the feedback loop do this job, emphasizing or attenuating determined frequency regions.

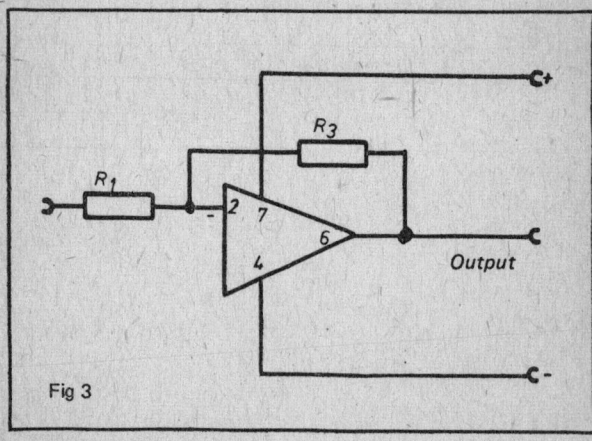

Fig 3

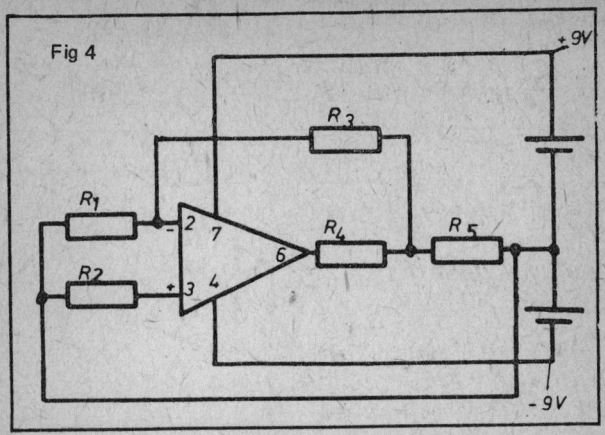

Fig 4

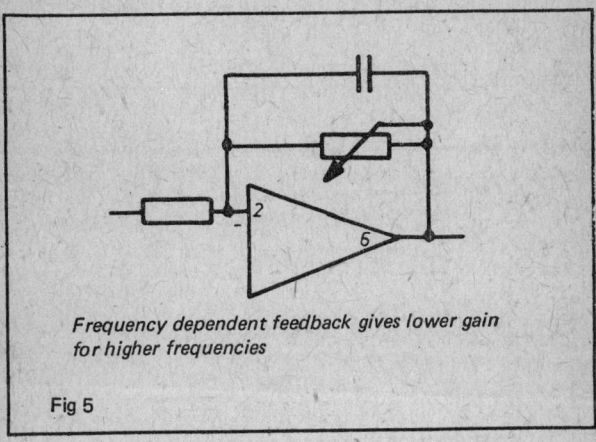

Frequency dependent feedback gives lower gain for higher frequencies

Fig 5

11

Figure 6 shows how high selective filters can be designed for selection of frequencies in remote control tone decoders for instance, by means of a T or twin Γ network in the feedback.

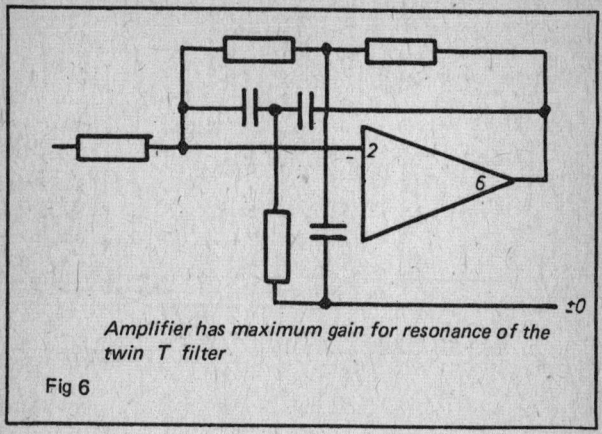

Amplifier has maximum gain for resonance of the twin T filter

Fig 6

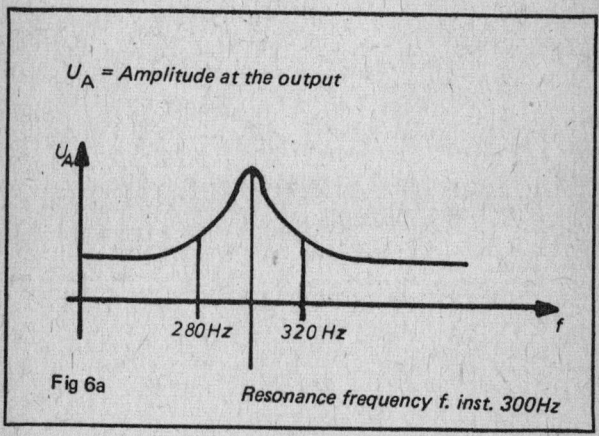

U_A = Amplitude at the output

Fig 6a

Resonance frequency f. inst. 300Hz

280 Hz   320 Hz

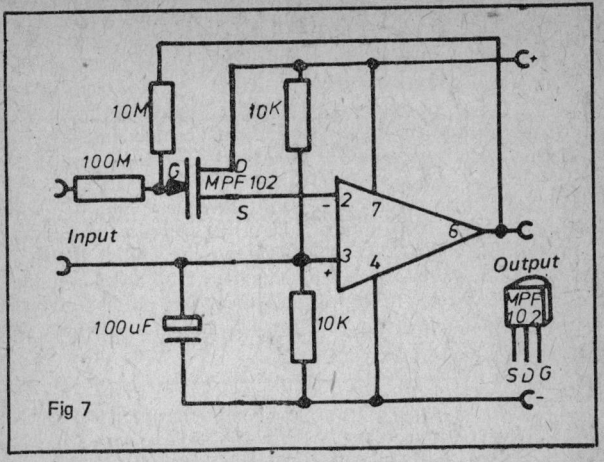

Fig 7

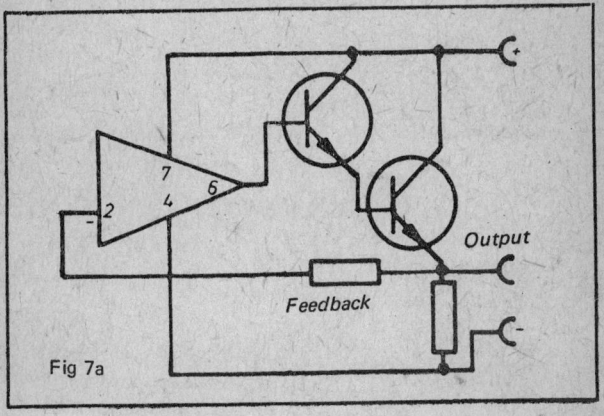

Fig 7a

13

INCREASE OF INPUT RESISTANCE

The input resistance can be increased by a Mos Fet transistor to some Gigaohms. The output current capability too can be increased by a transistor stage, for power supplies for example. The following circuits are only a few of the number of possible solutions and the amateur will find more applications of this cheap allround component very soon, after gathering some experience with the given circuits.

The described component can be used for frequencies up to 3 MHz, other amplifiers, the μA 702 from Fairchild or the equivalent component from Motorola, the MC 1712 CG operate up to 65 MHz. It should be noted, that the pin diagram of this amplifiers is different from that of the type 741. Except for the different frequency compensation and the non symmetrical supply voltages all comments of this book concern the 702 also.

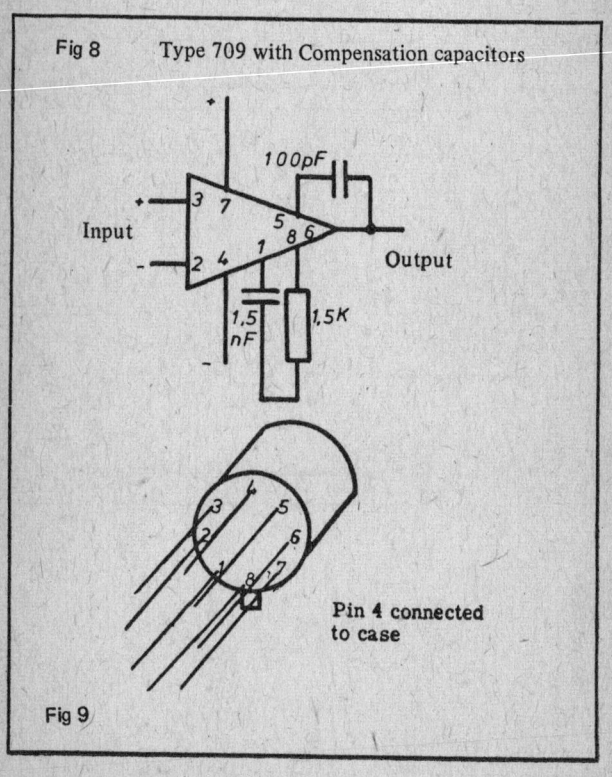

Fig 8   Type 709 with Compensation capacitors

Input

Output

100pF

1,5 nF

1,5K

Pin 4 connected to case

Fig 9

Type Table

Manufacturer	Type 709	Type 741	Type 748
Valvo	TAA522	TBA222	
Motorola	MC1709CG	MC1741CG	MC1748
Fairchild	uA709	uA741	uA748
Siemens		TBA221	
Texas Instr.	SM72709	SM72714	
Tandy-Archer	RS276-017	RS276-010/741/007	

Typical Performance	709	741	748
u max	+18–18V	+18–18V	+18–18V
u supply	+9 to 15 −9 to −15V	+6 to 15 −6 to −15V	+3 to 15 −3 to −15V
R input	250KΩ	1MΩ	2MΩ
I output	10mA	10mA	10mA
Short circuit protected	No	Yes up to ±15V	Yes
Maximum frequency	1000KHz	1000KHz	1000KHz
External components required	Yes	No	No
Zero point adjustable	With restrictions	Yes	Yes

Lever components with the same characteristics are available in a dual inline package

Manufacturer	Dual 709	Dual 741
Motorola	MC1437L	MC1458
Fairchild	MA1437	MA747
National Semi.		1458
Tandy-Archer		RS276-038

PIN CONFIGURATION AND EXTERNAL COMPONENTS

The difference of prices between the 709 and 741 circuit is only a few pence but on the other hand two capacitors and a resistor are needed to compensate the 709.

The gain of the 741 is twice as high and the amplifier can be used in every circuit designed for the 709. The 748 meets higher requirements but is more expensive too.

Before beginning with the discussion of practical circuits, I will call to your mind that the + and − sign at the left side of the triangle do not mark the supply voltage pins but the non inverted and inverted input. The pin configuration is compatible for the three types and for all producers.

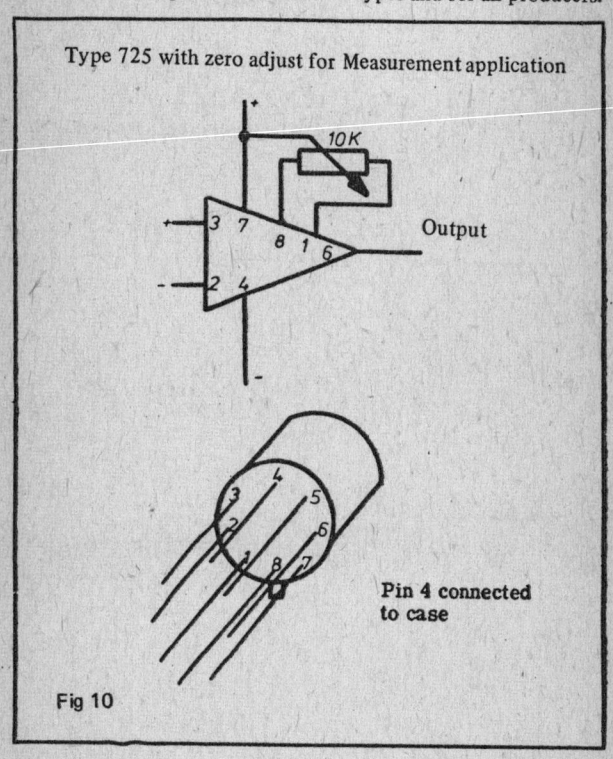

Type 725 with zero adjust for Measurement application

Pin 4 connected to case

Fig 10

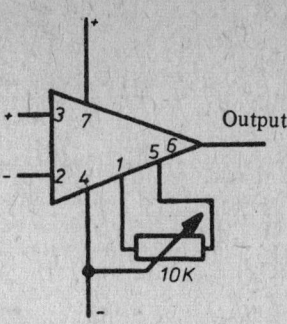

Type 741 with zero adjust for Measurement application

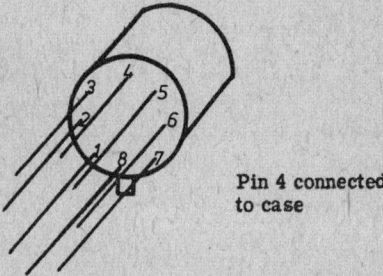

Pin 4 connected
to case

Fig 11

PREAMPLIFIER FOR CRYSTALPICKUP

Crystalpickups have a high source impedance and cannot be matched to low input impedance amplifiers. In this case it is necessary to lower the impedance to avoid loss of bass and also to amplify the power produced by the pickup. By varying the 100 kΩ resistor in the feedback loop the desired gain may be obtained. The amplification will be reduced with decrease of the resistor value.

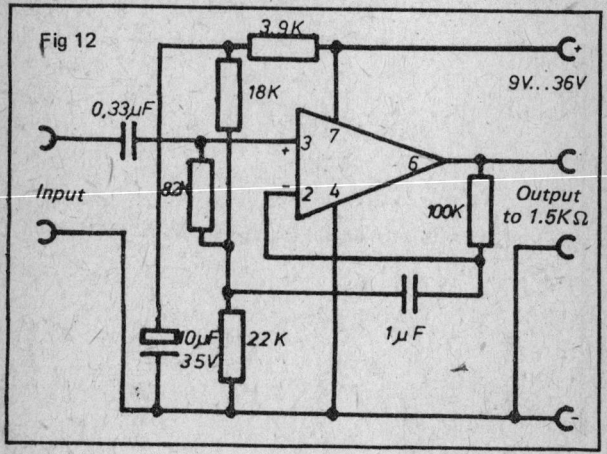

Fig 12

PREAMPLIFIER FOR MAGNETIC MICROPHONES

Figure 13 shows a preamplifier sufficiently sensitive for magnetic microphones with an output impedance of 1kΩ. As the circuit has a very high gain, it should be placed in a metal case, if not, line rippel or disturbances by transmitters will occur.

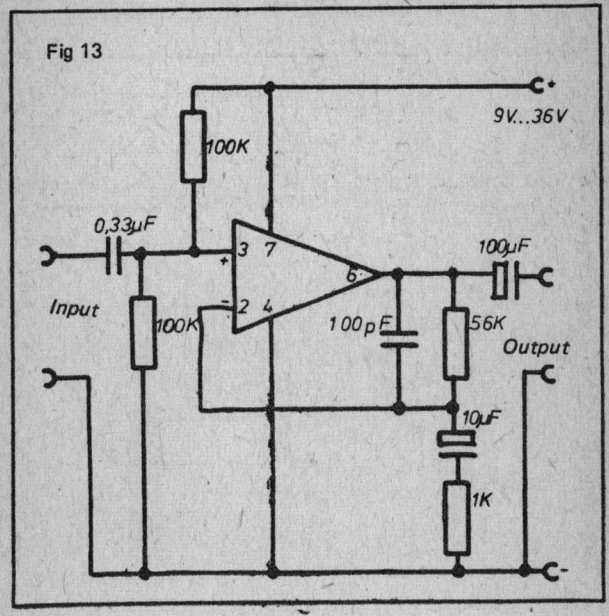

PREAMPLIFIER FOR MAGNETIC PICKUPS

High fidelity phonographs have magnetic pickups. In many amplifiers no appropriate frequency equalization is provided. By using the circuit 14 a suitable preamplifier may be constructed without difficulty.

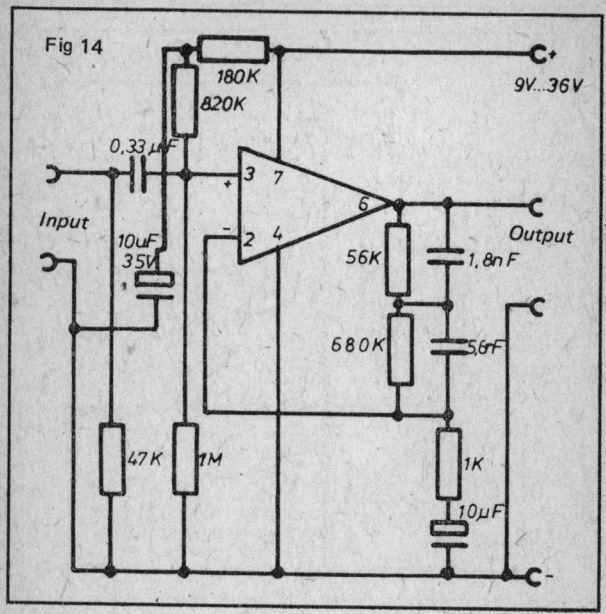

MIXING AMPLIFIERS

Audio mixers are affected with voltage loss because of the series resistors needed for decoupling the different channels. This loss must be compensated by linear amplification. The gain can be adjusted to the desired value by variation of the feedback loop.

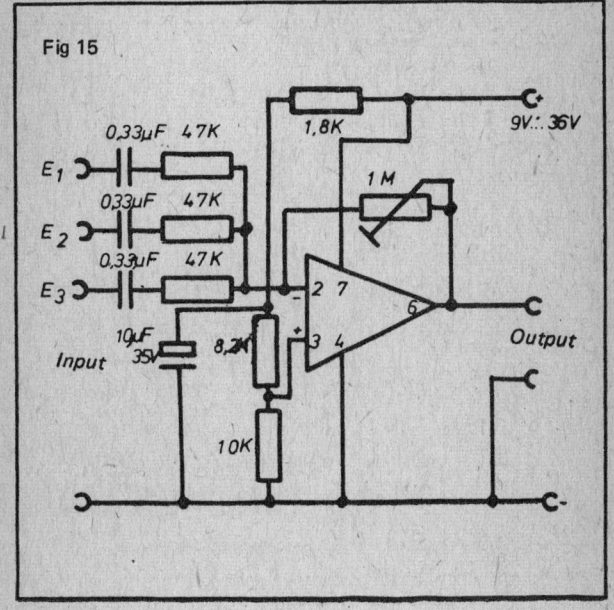

Fig 15

RECORD AMPLIFIERS WITH BASS BOOST

While they are playing, magnetic tape recorders cause signals
which are frequency dependent. The equalization amplifier shown
in Figure 16 will do this job.

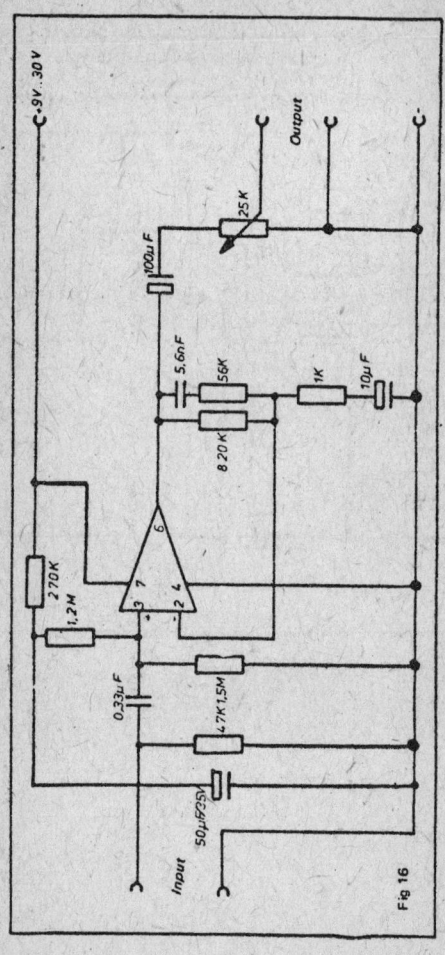

Fig 16

VOLTAGE FOLLOWER

The circuit shown in Figure 17 will be needed to match high impedance signal sources to low input impedance stages. The input and output voltage are the same. The input resistance amounts to 1MΩ whereas output impedance is not higher than 1kΩ.

The input signal (peak to peak) should not exceed 80 p.c. of the supply voltage. You should remember the fact that the peak to peak voltage is 2.8 times higher than the root mean square voltage.

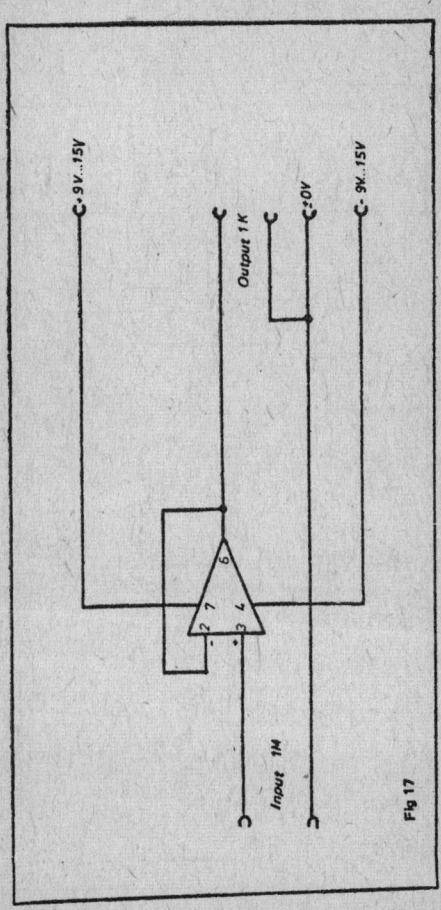

Fig 17

ACTIVE SOUND FILTERS

The passive sound filter may only be used to attenuate frequency ranges. The sound filter shown in Figure 18 however both attenuates and amplifies high and low frequencies. This means, that the control range is doubled. The sound controls consisting of T-filters are placed in the reverse feed-back loop and influence the gain-frequency.

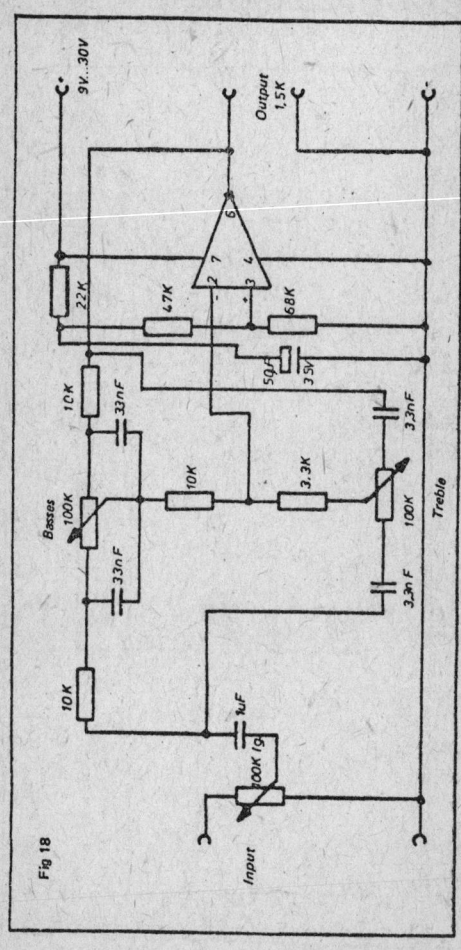

Fig 18

The measurements were made with a supply voltage of 30V. The input signal was 1V pp. Maximum input signal (30V supply voltage) is 2.5V pp. The harmonic distortion for 1 kHz is 0.03%. Noise voltage is 0.3 mV pp.

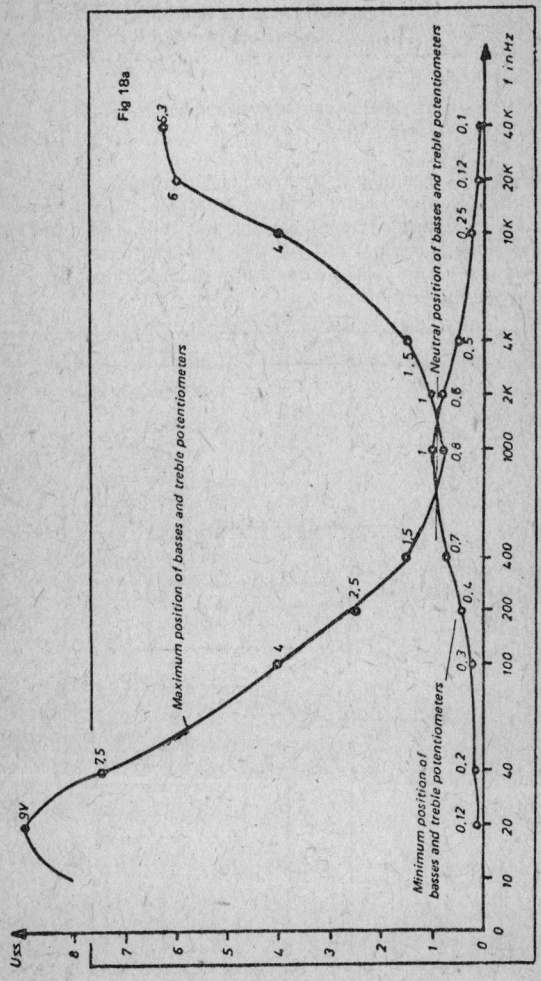

Fig 18a

BAND-PASS ACTIVE FILTER

A short time ago, the filtering of a desired frequency from mixed frequencies, e.g. music, tone signals of multi channel remote control and multi channel psychodelic light was only possible by using passive filters, built with R-C combinations or tuneable resonant circuits, consisting of coils and capacitors. With the help of integrated amplifiers, it is now possible to construct band-pass active filters of high quality and small size.

Figure 19 shows a circuit that can be designed for any frequency by varying the condensors.

With the help of a trimming resistor the transfer characteristic can be influenced. Selectivity reaches its maximum when the tolerances of the resistors and condensors are less than 1 p.c. You should take care that C_3 has double the value of C_1 or C_2. It is best to take four condensors of the same value, two of which should be used in parallel connection for C_3.

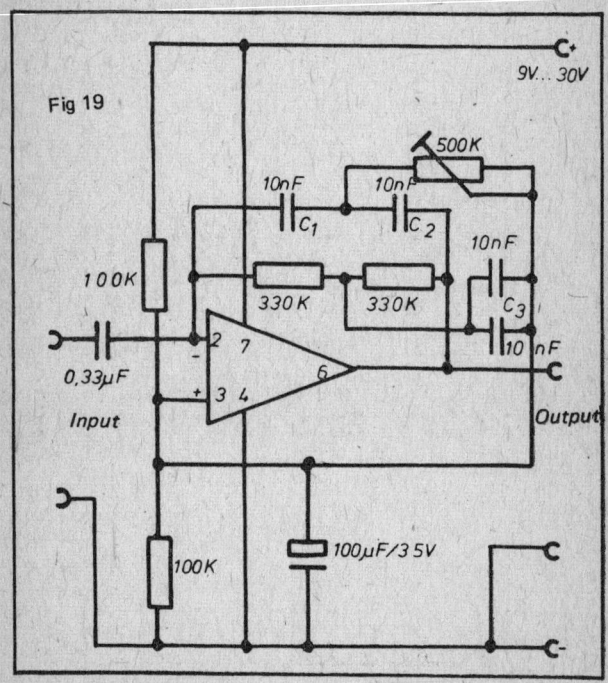

3 WATT MUSIC AMPLIFIER

It is very cheap and requires hardly any work to build a little high quality amplifier with 0.6V input sensitivity. The output transistors should be matched and have high current gain. If a higher input signal is available, the feedback impedance can be reduced, and thus the harmonic distortion will diminish. The harmonic distortion amounts to 0.3% for 1000 Hz in the design shown in Figure 20.

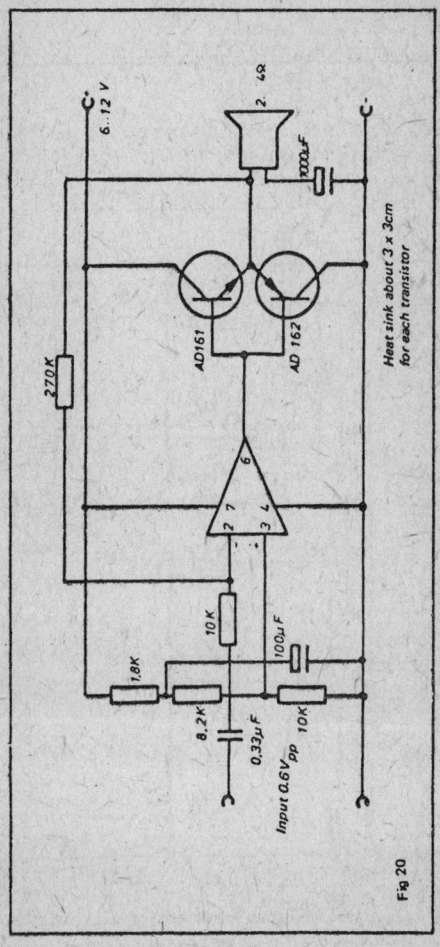

Fig 20

TELEPHONE MONITOR AMPLIFIER

Often it is necessary to let several people listen for a phone call. As direct connections are forbidden by the G.P.O. the weak magnetic fields of a telephone set can be made audible by means of an inductance coil and the amplifier shown in Figure 21. If the amplification of high frequencies exceeds normal values, the capacitor in the feedback loop may be increased to 270 pF. An inductance coil with rubber sucker is easily obtainable.

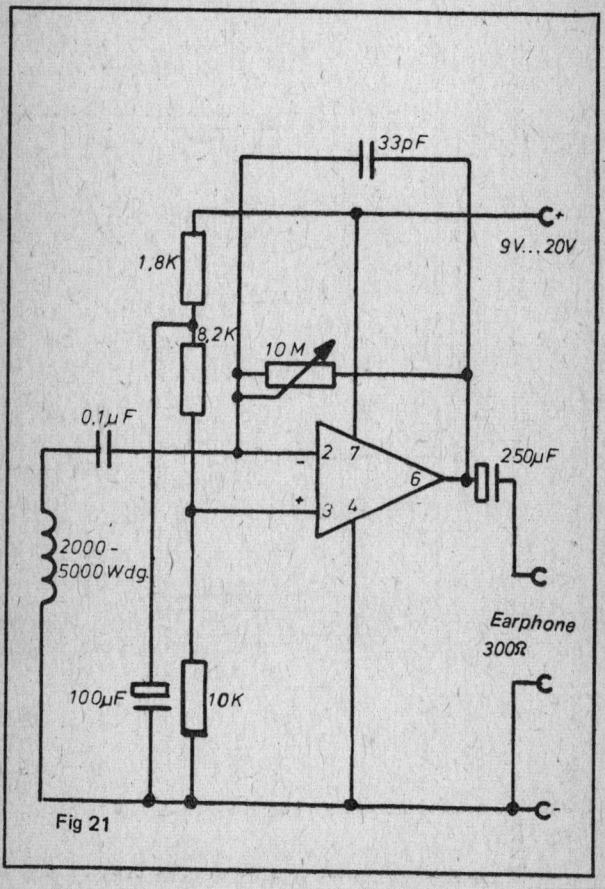

Fig 21

CABLE DETECTOR

The cable detector functions in the same way the circuit of Figure 21 does. A magnetic field must be generated to find the cable. The magnetic field depends on the current. Therefore one must connect a heater or some other heavy current consumer to the mains. If for example in a new house the mains has not yet been connected or there are bell, alarm or antenna circuits that may not be loaded with strong current, the multivibrator shown in Figure 23, should be connected to the wiring you wish to trace. A frequency of 400 to 800 Hz should be used.

A relay coil, if possible with an U or E shaped core can be used as detector coil. The current consumption amounts up to 6 or 10 mA.

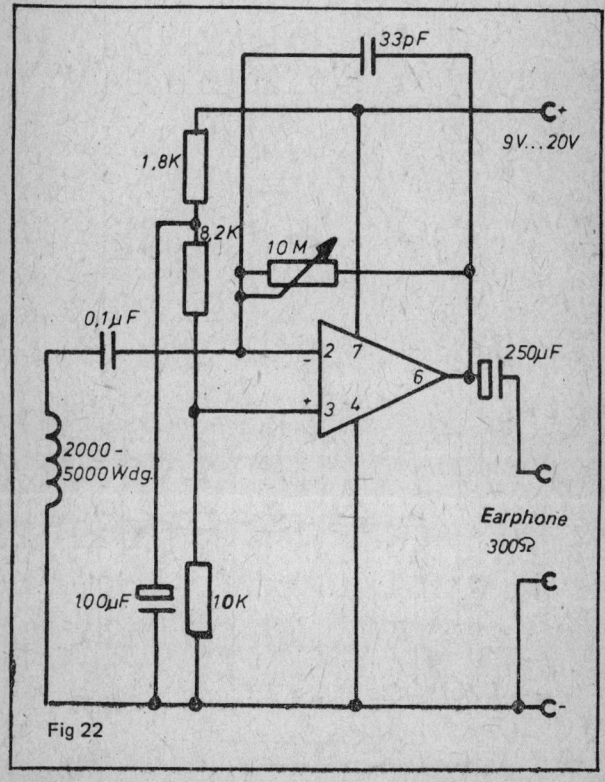

Fig 22

MULTIVIBRATOR FOR SYMMETRICAL SQUARE WAVE

Many multivibrator circuits are available but symmetrical square wave with steep pulse slopes is obtained only with specially designed circuits. In the circuit shown in Figure 23 time period is determined by only 1 capacitor and 1 resistor. Because of the high gain the pulse slopes are very steep. The coarse frequency regulation is by variation of the capacitor and fine regulation by adjusting the 2.5 MΩ potentiometer. The maximum frequency range : 0,1 to 20 000 Hz. The output voltage amounts to about 80% of the supply voltage, the slope time is 30 μsec for ± 5V supply voltage.

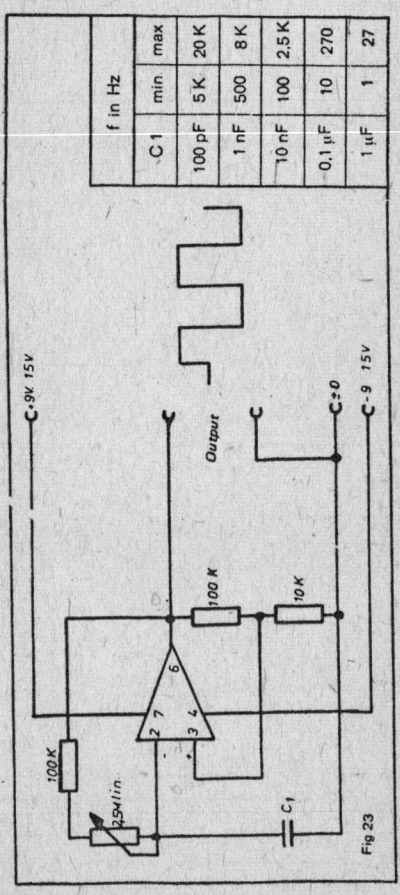

f in Hz	min	max
C 1		
100 pF	5 K	20 K
1 nF	500	8 K
10 nF	100	2.5 K
0.1 µF	10	270
1 µF	1	27

Fig 23

MULTIVIBRATOR FOR NONSYMMETRICAL SQUARE WAVE

The circuit of Figure 23 supplies you with a symmetrical square wave. For many purposes however a different duty cycle is needed. Except for the 5 MΩ trimming resistor, the circuit is the same as the one shown in Figure 23. Instead of taking a 2,5 MΩ trimming resistor or potentiometer, two pairs of resistors or potentiometers and diodes are connected in series as shown in Figure 24. This circuit gives you different charging and discharing times for different resistor values.

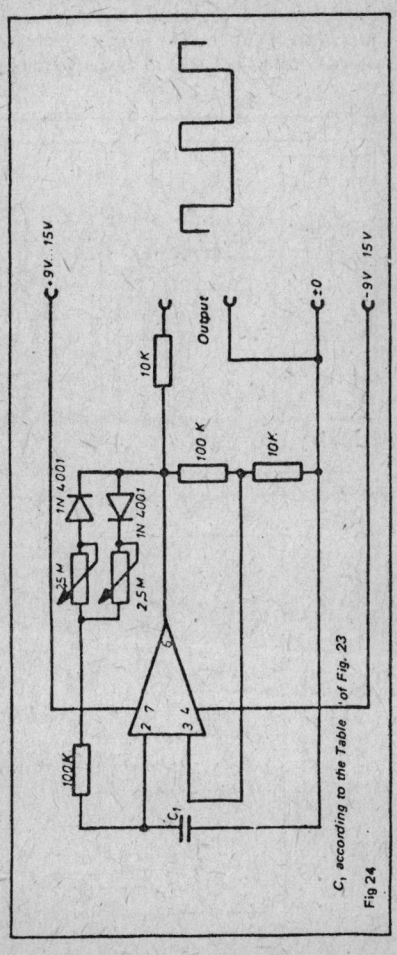

Fig 24

NEEDLEPULSE GENERATOR FOR NEGATIVE AND POSITIVE PULSES

Many experiments with modern components require a needlepulse generator, from which positive or negative pulses can be taken. The multivibrator described in Figure 23 is employed here to produce a square wave voltage. This is differentiated and devided into negative and positive pulses by diodes. The amplitude can be limited with Z-diodes. The duration of the pulses you produce may be changed by variation of the capacitor C_2. The steep pulse slope is sufficient to drive modern integrated counters. The output which is not needed must be connected directly to earch (± 0V) to discharge the capacitor C_2.

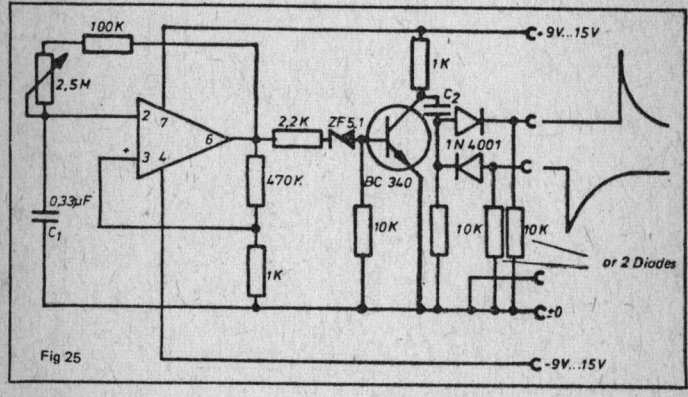

Fig 25

C_2	t in µsec
100 pF	0,1
1 nF	1
10 nF	10

WIEBRIDGE WITH IC

For a long time Wienbridges have been used to produce sine wave voltage with variable frequency. With the help of the IC the circuit may easily be constructed. The setting of the trimming resistor R_8 and the stabilisation of the supply voltage are a little critical. With the help of an oscilloscope, connected to the output, the trimming resistor may be adjusted so that you get sine wave voltage at the output. As the load must be constant, it is necessary to connect an amplifier to the output. If the feedback of the amplifier decreases, first a trapezium wave voltage is produced and after continued increase of the feedback resistor a square wave voltage will appear.

The frequency range is from ahout 0,3 to 50 000 Hz and is very suitable for the testing of Hi-Fi-amplifiers and sound filters.

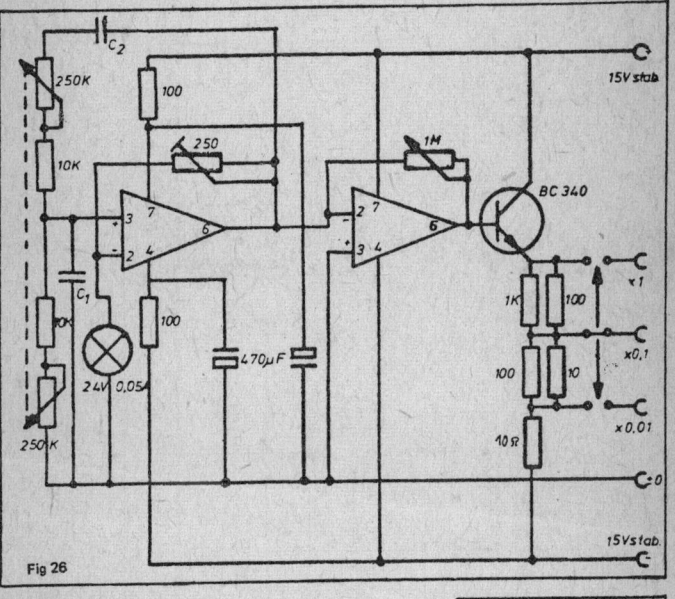

Fig 26

C 1, C 2	max.	min.
100 pF	50 kHz	5 kHz
1'nF	15 kHz	500 Hz
10 nF	1,5 kHz	60 Hz
100 nF	150 Hz	7 Hz
1 µF	7 Hz	0,3 Hz

MONOSTABLE MULTIVIBRATOR

If one needs to stretch short needle pulses the circuit shown in Figure 27 may be used. The differentiation of the output manages to delay the pulses. The relation of time is found with the formula: t = R . C The unit for C is Farad and for R Ohms. The fine adjustment is done by the trimming resistor 500 kΩ.

R	C	t_{msec}
0,5 M	1 nF	0,075
0,5 M	10 nF	0,5
0,5 M	0,1 µF	5
0,5 M	1 µF	50
0,5 M	10 µF	500

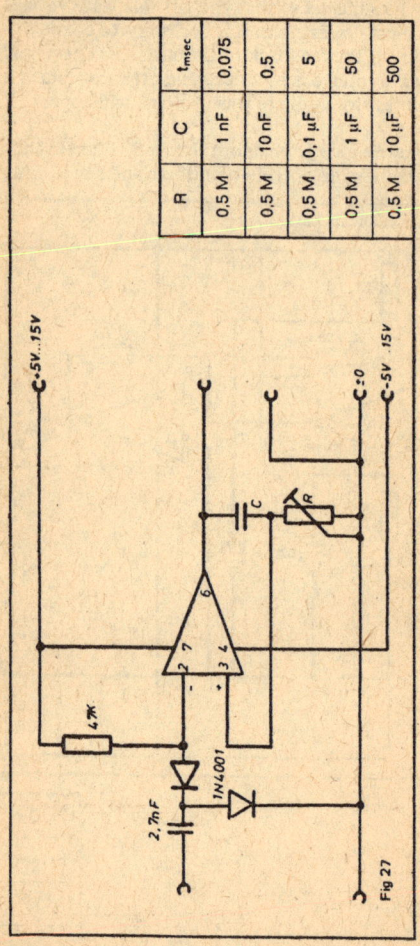

Fig 27

FLIP-FLOP WITH IC 741

There are many Flip-Flops or bistable multivibrators on the market to choose from, some of which are available as integrated circuits. The circuit shown in Figure 28 however makes some applications possible which are not easily attained with other circuits. If you make the input negative the output goes to positive voltage. Only a positive pulse causes the output to go negative, one or even several negative pulses do not. As the circuit reacts only when pulses are of the appropriate polarity, conditions with a high noise margin may be obtained. The circuit is suitable, for example, to reset a sawtooth generator in an oscilloscope by the negative part of the sawtooth. The trigger pulse for switching on again is then given by the pulse of the Y-amplifier. If the polarity of the reset pulse should be varied, a second input can be installed with the help of a capacitor at the pin 3 of the 741.

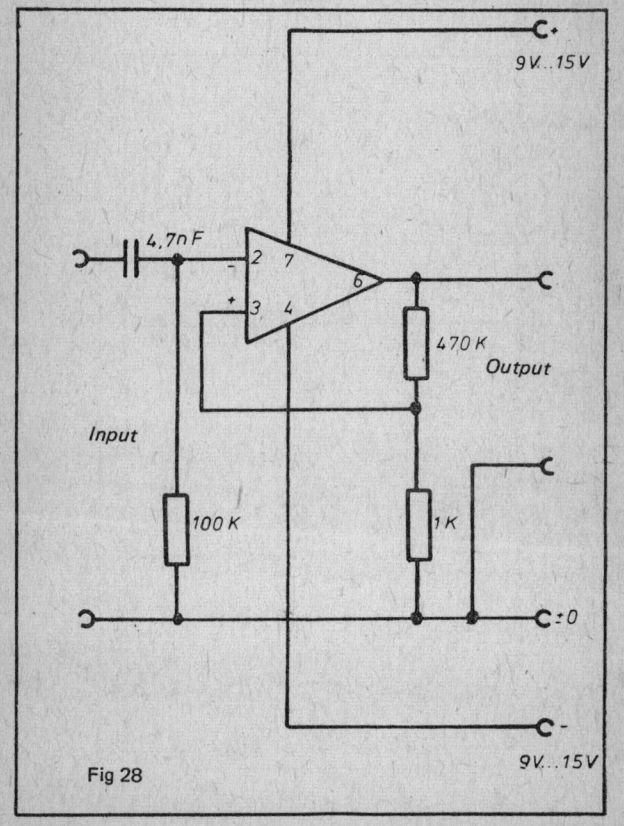

Fig 28

DELAYING LIGHT LEVEL DETECTOR

Weak changes of light intensity should not cause a light level detector to turn the connected units on and off. Therefore it must react slowly. The first component functions as an integrator, that means the average signal of light intensity change for a time period appears at the output. The second amplifier is connected in series so that the transistor may be opened in a short time.

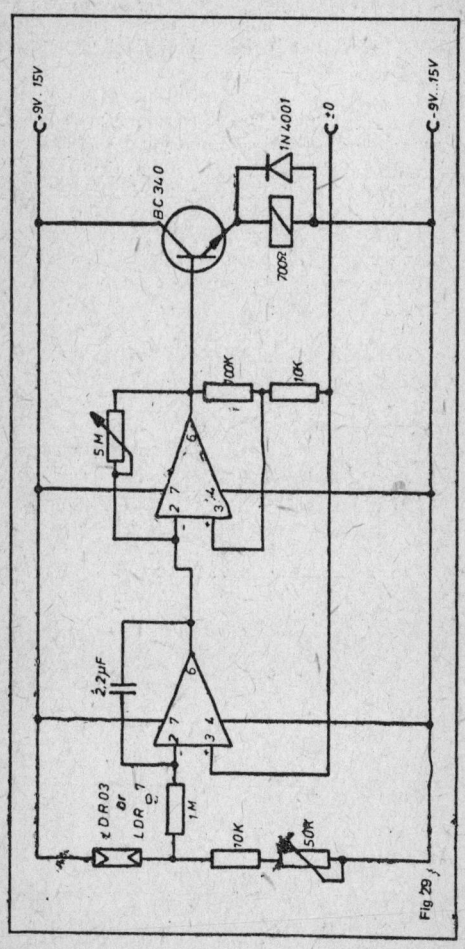

Fig 29

TIMER

Timers with a long pulse duration function well when they are constructed with MOSFETs. As the gate has an insulation resistance of 10^{16} up to 10^{18}, it is possible with small capacitor and very high resistor values to attain time periods up to 15 days. Automatic fish feeding machines or flower watering machines may be constructed, for example, with this circuit. Short time periods are suitable for the exposure of films or of photographic plates for printing circuits. The provided time periods may be found with the formula $C \cdot R = t$ in sec. The dimension of C is Farad and of R Ohms.

When the button "Ta" is depressed, the capacitor C_1 will be discharged. The gate of the MOSFET is then made negative and the MOSFET is blocked. After a time, the gate becomes more positive, the MOSFET conducts and the amplifier at output 6 goes negative. The transistor BC 360 opens and the relay closes. The transient suppression diode, in parallel connection to the relay, shortens inductive voltage peaks when the relay is turned off. The diode before the MOSFET assures that no positive voltage reaches the gate, otherwise the MOSFET would be destroyed.

If instead of the BC 360 a NPN-transistor is built in, e.g. a BC 340, a converse of the function takes place (dotted drawing). As the circuit has high impedances, a perfect isolation of the MOSFET, the capacitor and the button must be assured. In case the circuit doesn't function well you should try to lower the value of the resistor 1 MΩ and the potentiometer 5 MΩ. If the circuit functions now, the isolation is insufficient. Results were very good when the critical components were connected by free wiring unpolarised. During a test time periods up to 15 days were obtained. Suitable integrated amplifiers are the 709, 741 and 748. Any silicon transistor of at least 500 mA collector current and 40V maximum voltage can be used as transistor.

The timing capacitor must have a high insulation resistance.

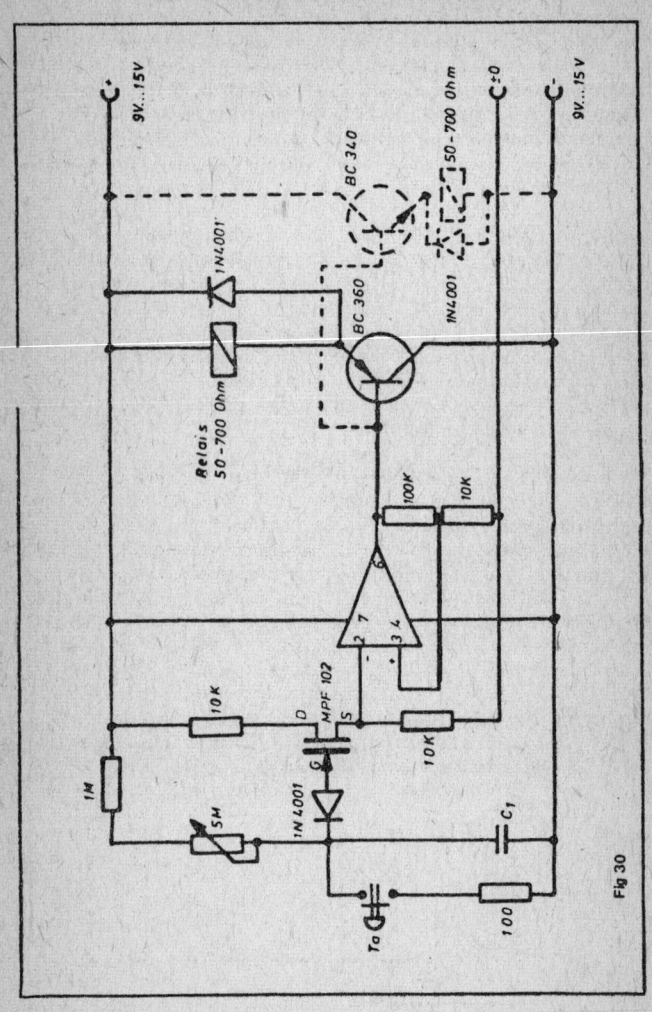

Fig 30

SWITCHER FOR FAST MOTION GENERATOR

Amateur photographers may perform fast motion shootings with a camera and the generator shown in Figure 31. The only difference to the circuit 29 being in the operating of the relay contact which discharges the timing capacitor. The resulting short pulse triggers the camera to go 1 picture further. The electrolytic capacitor in parallel connection with the relay may be enlarged or reduced according to the type of camera being used.

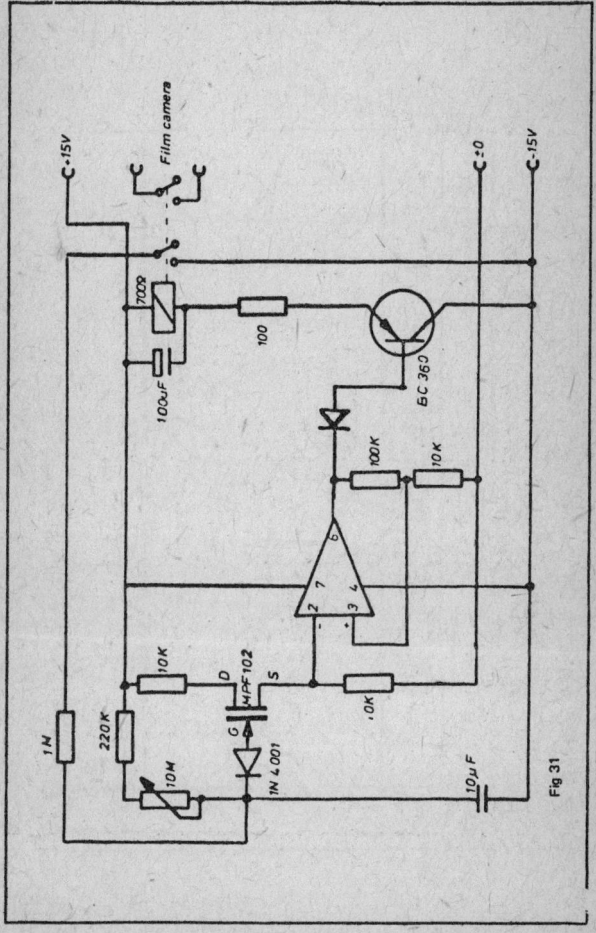

Fig 31

LIGHT SWITCH WITH IC

As the components have two inputs, it is possible to construct a bridge circuit of high sensitivity. When a change of light intensity varies the resistance of the photoresistor, the balance of the bridge is disturbed and the output transistor is opendd. No Schmitt trigger is needed because of the high amplification (Figure 32).

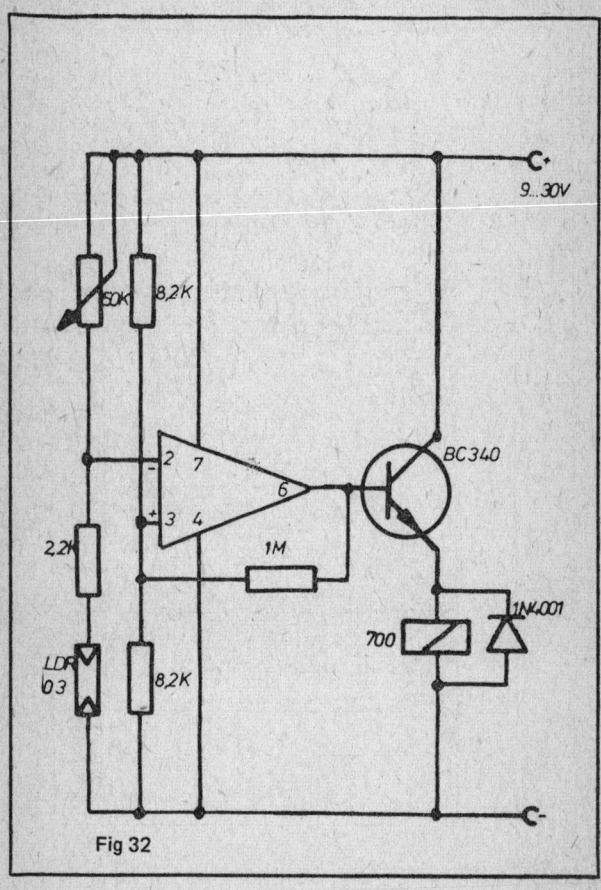

Fig 32

SIMPLE ANALOGUE TO DIGITAL CONVERTER

The popularity of digital instruments is increasing constantly, but in general these instruments are too expensive for amateurs. A self made electronic counter constructed with modern ICs costs less. For this purpose, the analogue signal dc voltage must be converted to pulses.

Circuit number 33 shows a simple analogue to digital converter with a precision of about 2%.

The input voltage produces at the output 6'of the 741 first a time delayed slope, then the unijunction transistor D 13T1 which can be programmed fires and discharges the capacitor 47 nF in the feedback loop. So a sawtooth voltage is produced with different frequencies depending on the input voltage. When the input is lowered, the potentiometer must be adjusted so that no pulses are produced. The calibration at e.g. 10,000 Hz for 1V is done with the potentiometer 10 kΩ.

In case the unijunction transistor is not easily obtainable, it can be simulated by two complementary transistors.

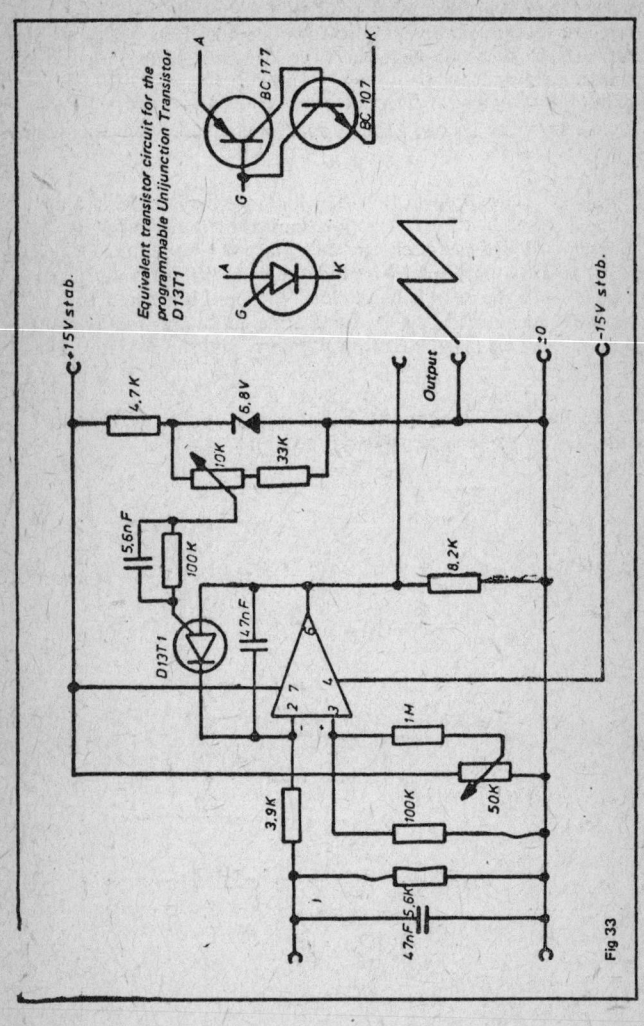

Equivalent transistor circuit for the programmable Unijunction Transistor D13T1

Fig 33

POWER SUPPLY FOR NEGATIVE AND POSITIVE VOLTAGES

For some apparatus it is necessary to have a voltage that can be regulated continuously between the voltage maximum, zero and negative voltage. It is possible for example to drive a model train forward and, without switching, backward too. As the power dissipation in the output transistors are converted to heat, the transistors have to be placed on appropriate heat sinks. ($R_{th} \leqslant 3,5$ W)

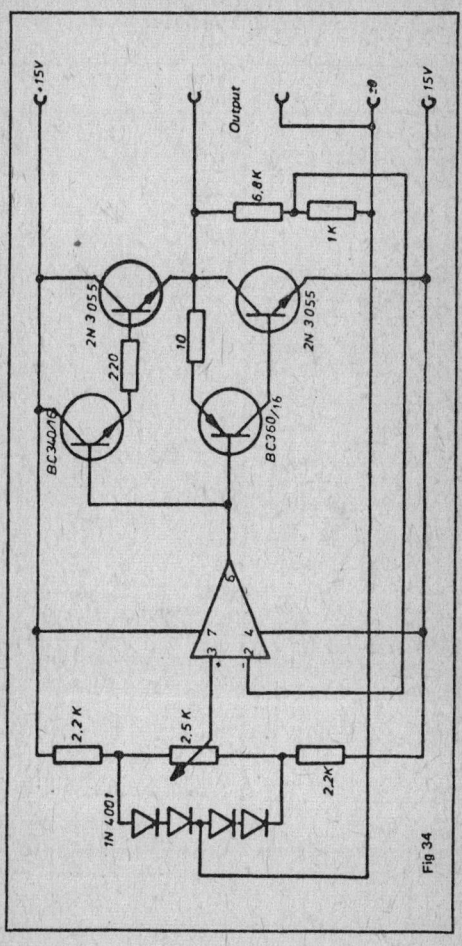

Fig 34

SCHMITT-TRIGGER

Figure 35 shows a Schmitt trigger with two inputs which can be triggered by a negative or positive input voltage corresponding to the input you use. The adjustment of the 10K potentiometer determines whether the output goes negative or positive without an input signal. The hysteresis too depends on the position of the potentiometer. The voltage comparator output changes if the input exceeds or falls short of half the supply voltage. The risetime of the output signal amounts to 30 sec for 10V supply voltage, the input resistance is about 100 kΩ.

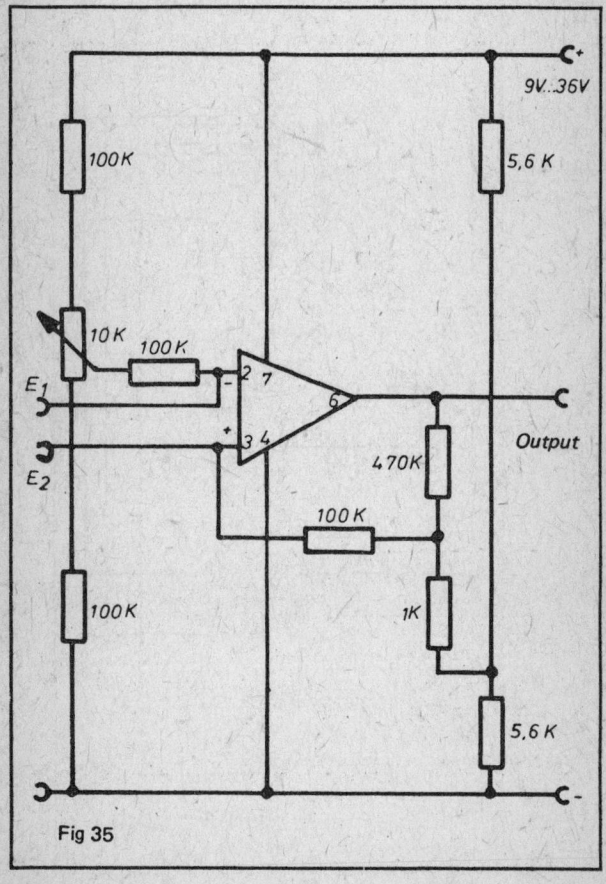

Fig 35

NANOAMPEREMETER FOR DC CURRENT

If a weak current flows from 3 to 2 or reverse, the input resistance produces a voltage difference which causes the output to go respectively or negative. So the meter indicates currents in the nanoampere range. The instrument is calibrated when several resistors which decrease with powers of ten, e.g. 100 Ω, 10 Ω, 1 Ω, 0.1 Ω are placed between 2 and 3. Now a higher current which can be indicated on a normal meter is fed to the circuit. The smallest resistor should be tried first. If for a current from A to B of e.g. 1 mA the meter indicates exactly 1 mA, it can be concluded that for a shunt resistor of 100 Ω, 1 microamp is sufficient to get a meter deflection equal to 1 mA.

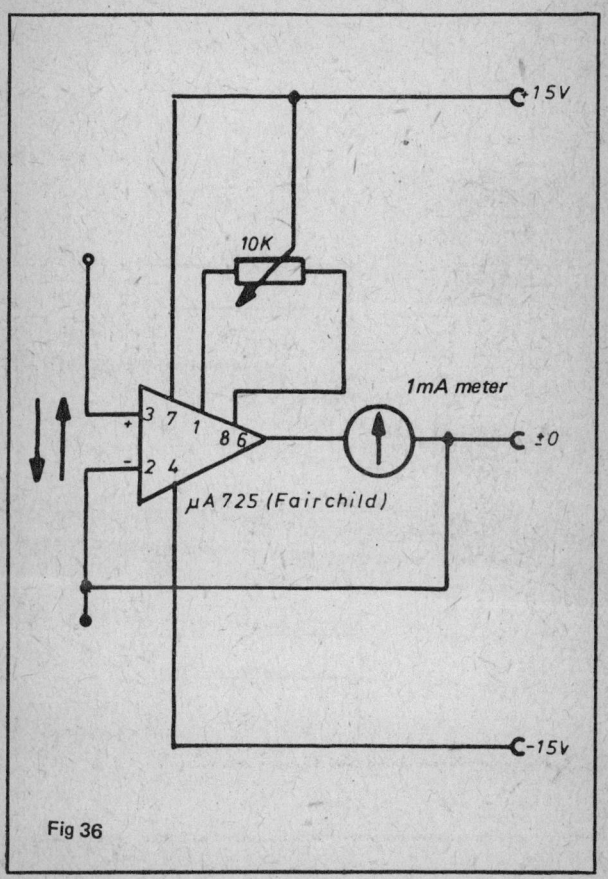

Fig 36

MULTIVOLTMETER AND MICROAMPEREMETER WITH PROTECTED INPUT

It is possible with the help of the circuit shown in Figure 37 to construct a high sensitive multimeter without difficulty. The temperature drift is about zero and the instrument contrary to valve voltmeters, after switching on is instantly ready for service and stable in operation.

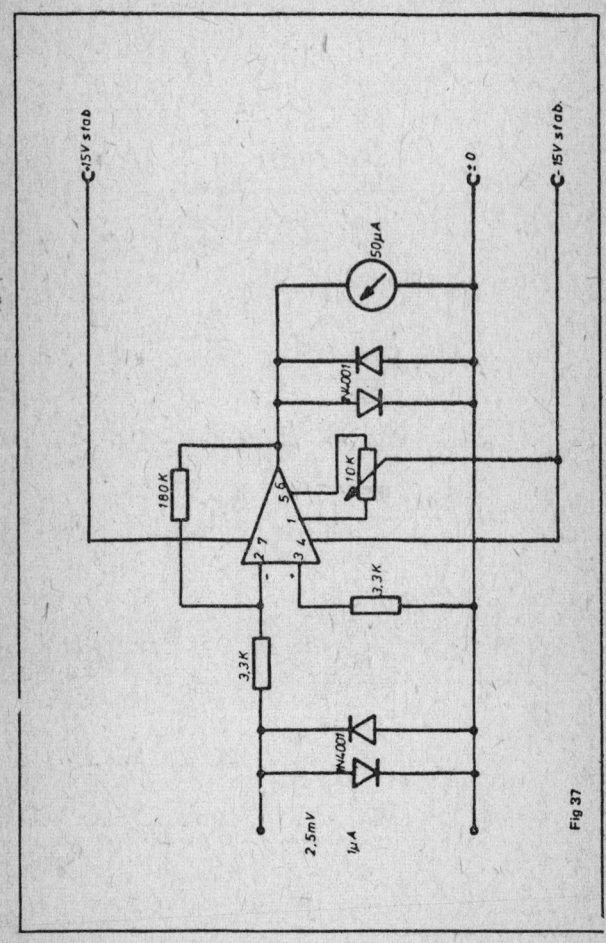

Fig 37

ALTERNATING CURRENT VOLTMETER WITH LINEAR SCALE

The construction of multimeter for ac voltage and current is rarely tackled by amateurs because the scale of the meter is not linear and is difficult to draw. The circuit 38 however presents an ideal rectifier, the scale of which is perfectly linear. With the help of suitable switches it is now possible to combine circuits 37 and 38 so that a precision multimeter can be constructed.

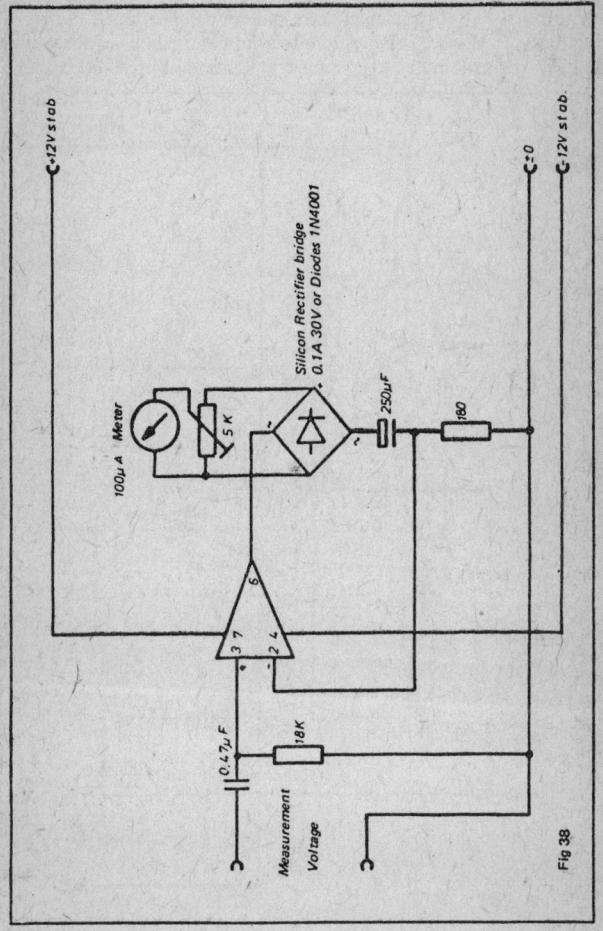

Fig 38

REMOTE THERMOMETER

Germanium transistors have an easily measurable C-E-leakage current depending linearly on temperature. The resistance variation is higher than that of NTC or PTC-resistors. With the zero point potentiometer the index is first brought to zero position. To achieve this the germanium transistor has to be placed into an ice water bath, as it is well known the temperature of the water is then exactly 0°C. Hereafter the ice will be removed and the water will be warmed slowly. To control the effect a normal thermometer can be dipped in the waterbath. In this way the remote thermometer can be calibrated with the 250K potentiometer to a water temperature of up to 50°C.

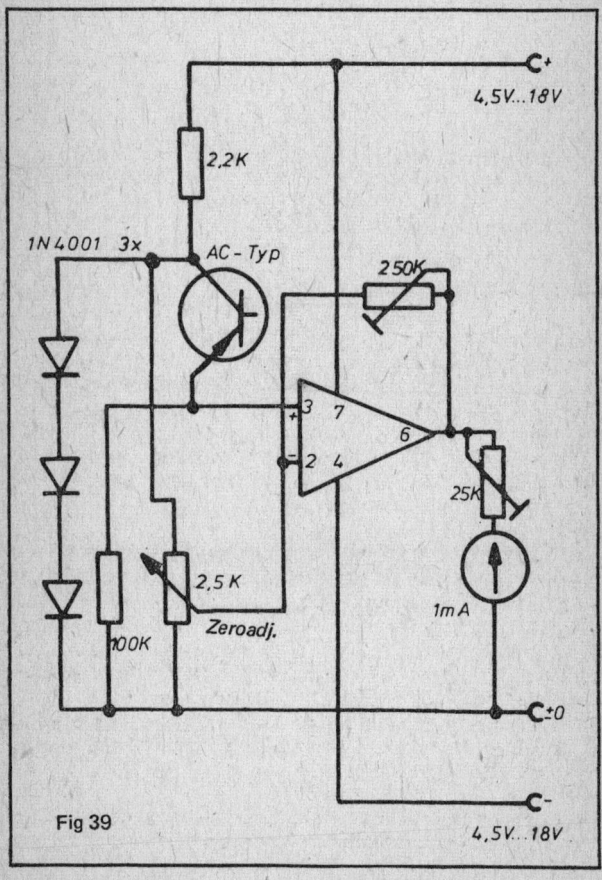

Fig 39

STABILIZED VOLTAGE SUPPLY FOR SINUSOIDAL AC VOLTAGE

Stabilized dc voltage has become a matter of course for electronics. For several decades now ac voltages have been stabilized rather well with complex stabilisation circuitry. It is however very difficult for amateurs to design and construct such circuits and often the output waveform is that of an extremely distorted sine wave. The circuit shown in Figure 40 guarantees high stabilisation with a voltage that can be regulated. The sine wave shape is nearly maintained.

The unit consists mainly of two transformers which are linked by a rectifier bridge. If the transistor connected to the dc poles turn off, the bridge does not conduct and no voltage is induced in the second transformer. The reverse extreme is a fully opened transistor. The bridge is now low ohmic and the second transistor is fed with full current. With the help of a differential amplifier the value can be varied between the two extremes.

As in the transistor 2N3055 the power dissipation is converted to heat, a sufficiently large heat sink must be used. The power of the circuit shown in Figure 40 may not exceed 70 VA. Otherwise several transistors are needed in parallel connection.

REGULATE STABILIZED VOLTAGE SUPPLY WITH HIGH POWER

The circuit shown in Figure 41 is suitable for speed control of motors using commutators or for continuous regulation of heaters. The trigger pulses for the thyristor must be synchronized with the mains frequency, then the voltage supply of the unijunction transistor causes synchronisation. The unijunction transistor is fed by D_1 with positive half waves which are limited by the Z diodes. During the zero crossing of the ac voltage needle pulses are produced that fire the UJ-transistor and discharge the capacitor C_2.

The rate of which the IC takes to charge C_2 and thus to fire the UJ is dependent on the difference across its input. The phase control trigger point is chosen automatically and the output voltage remains constant. With independently excited dc motors speed variation of 1 to 2% at 4000 r.p.m. can be achieved. The fire or trigger transformer can be made of a ring core of about 15 mm diameter. The choke coil Dr.1 smooths the dc current fed in by the thyristor. The thyristor 2N4443 by Motorola is only suitable for ohmic consumers. For inductive load the type 2N4444 should be used. The maximum current usable for both types is up to 8A. Other thyristors may also be used.

If operated from the 220 ac-Line, the break down voltage should amount to 400 V for ohmic load and to 500 V for inductive load.

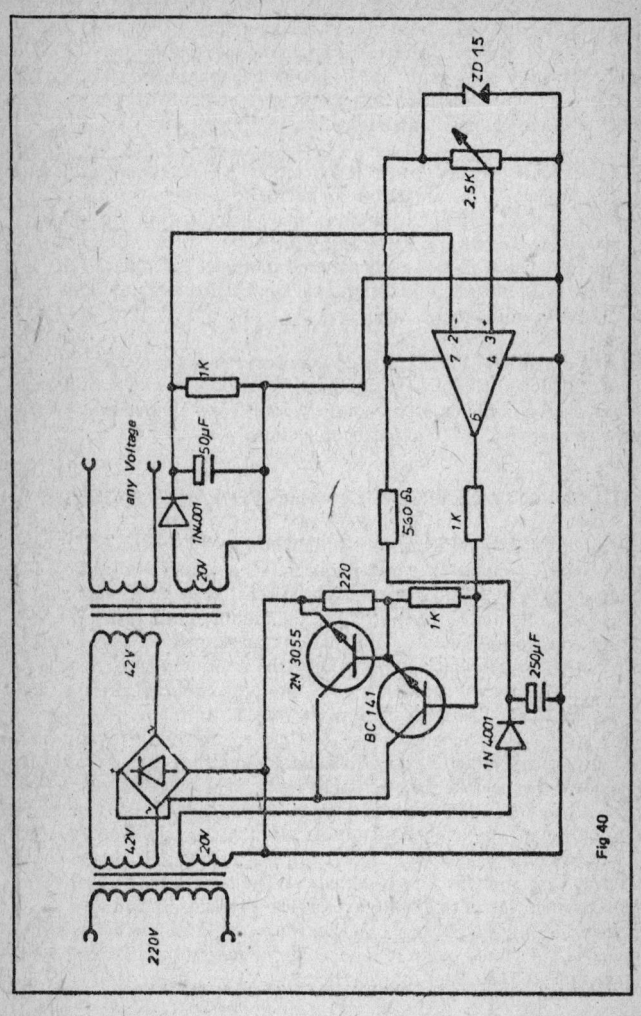

any Voltage

ZD 15

2.5 K

1K

50 µF

1N4001

20V

20V

42 V

530 Ω

1K

2N 3055

BC 141

220

1K

1N 4001

250 µF

42 V

42 V

20V

220V

Fig 40

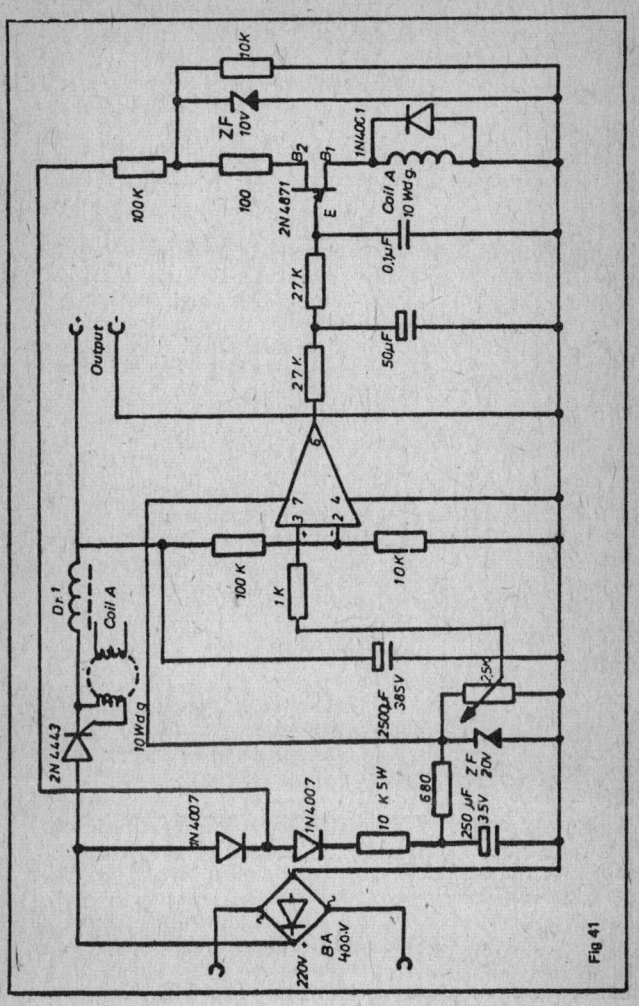

Fig 41

51

STABILISED VOLTAGE SUPPLY FOR 5V UP TO 60V, 5A
(Fig.42)

The differential amplifier is used here to regulate a dc power supply. The reference voltage is adjusted at the potentiometer P1. The output 6 goes negative and the transistor TR_3 doesn't conduct. Passing the resistor R_4 a current flows to the transistor Tr_1 and also Tr_2 which are both opened. At the output voltage divider a positive voltage is produced which then reaches the input 3.6 now goes positive as well and T_3 conducts. Part of the base current is therefore taken away from Tr_1 and the output voltage remains constant. Instead of the two transistors $Tr_1 + Tr_2$ a darlington transistor can be used. The power transistor must be adequately cooled.

The heat radiation reaches its maximum when the output voltage is low and the current high e.g. the heat radiation for 10 V output voltage 70 V input voltage and 2A load amounts to $60 \times 2 = 120W$. The 2N3055 has a maximum power dissipation of 70W with optimum cooling, thus the power limit is exceeded. If 60V for 5A are taken from the circuit the power dissipation amounts to no more than $70 - 60 = 10 \times 5 = 50W$.

STABILIZED VOLTAGE SUPPLY UP TO 30V and 5A
(Fig.43)

In this power supply the series transistors are directly controlled by by the IC. Other than this the circuit is the same as Fig. 42. It should be noted that the inputs are in reverse position, because contrary to the circuit 42 the transistor has no invert function.

VARIABLE REFERENCE VOLTAGE
(Fig.44)

Generally reference voltage sources produce stable voltages. The circuit shown in Figure 44 produces a reference voltage which can be regulated and remains constant between open output and a 10 mA load. The Z-diode can also have other voltage values, the temperature drift however reaches its minimum at about 5V.

If an absolute constancy of the voltage is required, the Z diode has to be replaced by reference diodes. A variation of the supply voltage modifies the current of the Z-diode and influecnes the voltage. With the help of a current stabilizing circuit, replacing the 3.9 KΩ resistor, this can be avoided (drawing on the above right). The constancy of the rest of the circuit depends primarily of the temperature characteristic of the Z-diode, the resistors and the potentiometer.

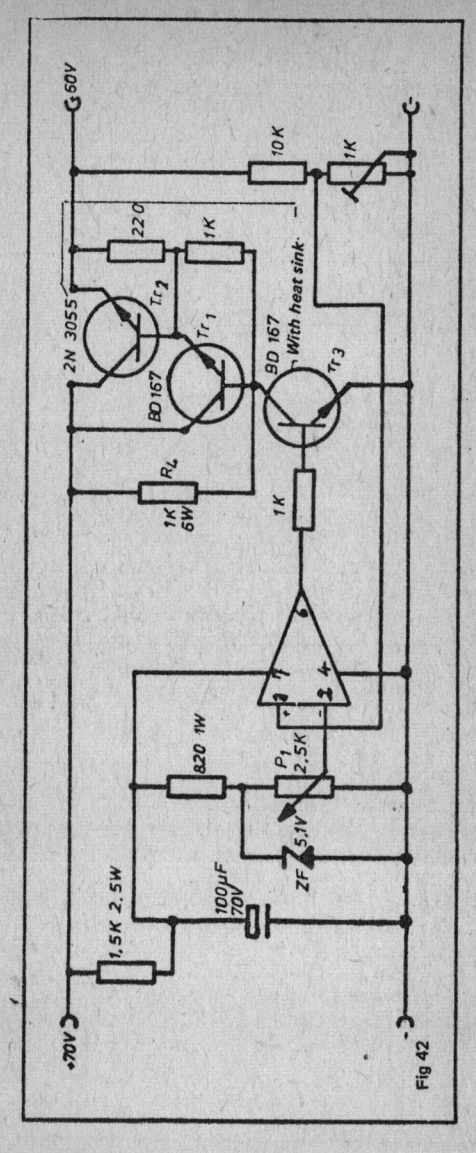

Fig 42

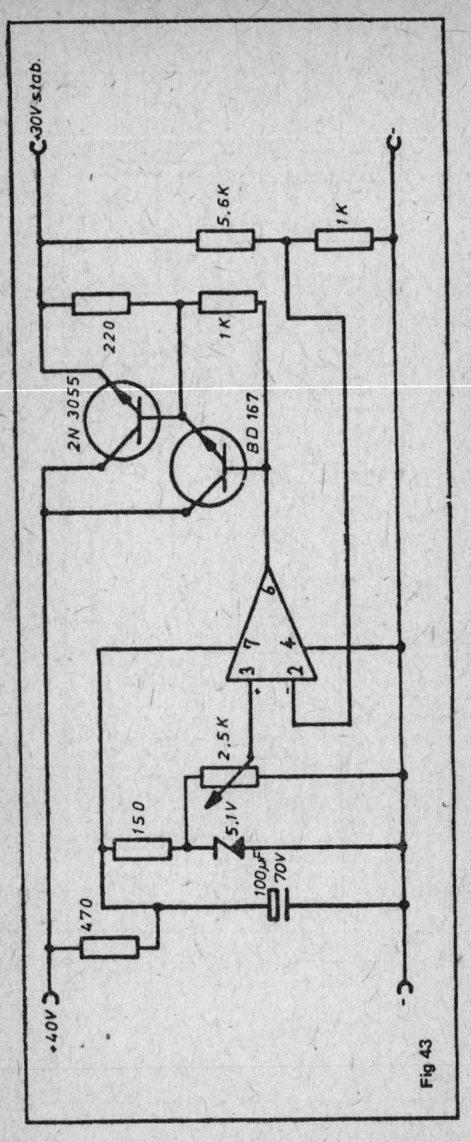

Fig 43

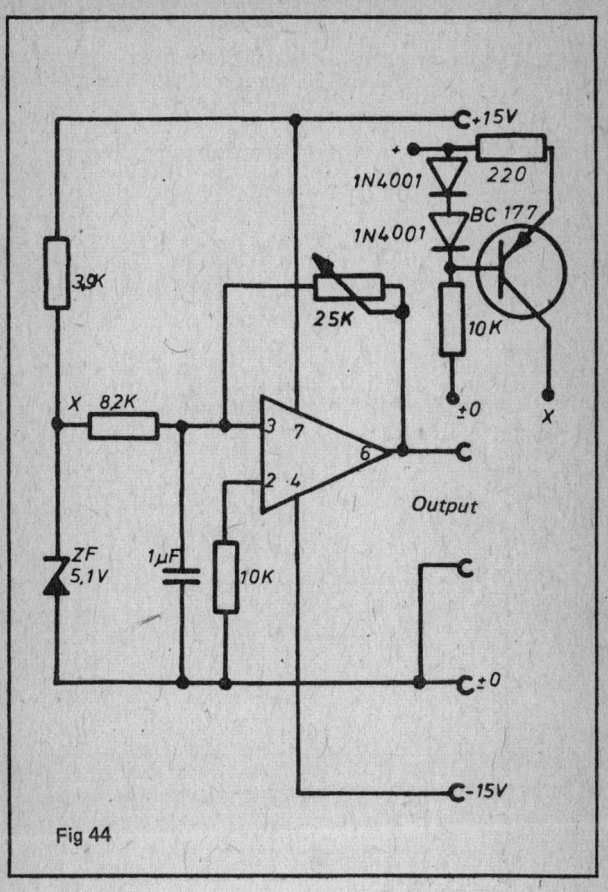

Fig 44

SERVODRIVER

This circuit is suitable for many purposes. A voltage is adjusted on the set point potentiometer. The motor then drives for example an antenna or a mixing valve, until the instantaneous value potentiometer has reached the same value. This potentiometer must be linked mechanically with the spindle of the gear motor. Other pick ups can be used instead of the set point potentiometer.

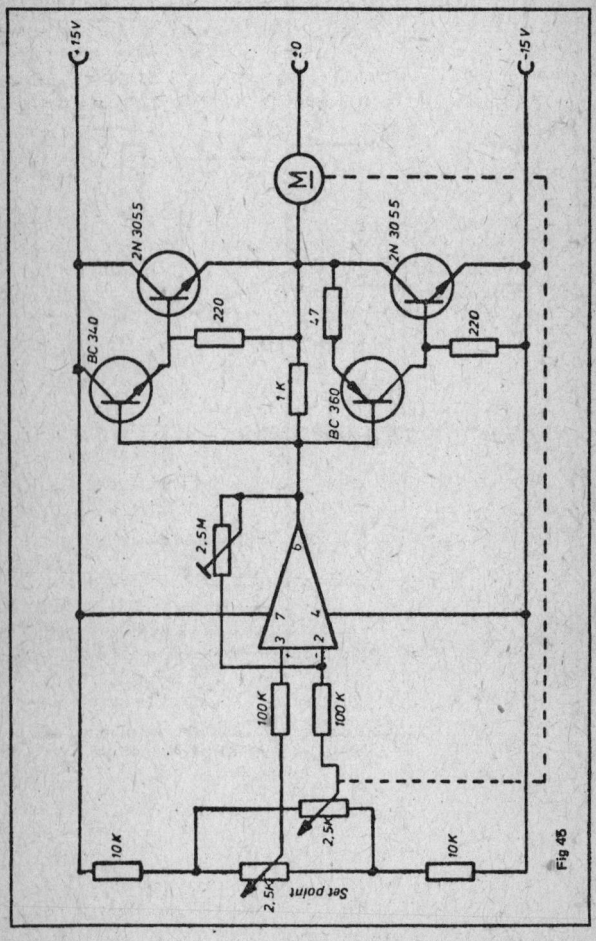

Fig 45

TALK BACK INTERCOM

In Figure 46 an intercom with high quality of speech is shown that may be built quite simply. The loudspeakers are used for listening or speaking depending on the position of the switches. For switching, relay contacts may be used. With the help of the 100 Ω potentiometers the volume can be adjusted. The preamplifier consists of one IC with ohmic (390 Ω) and a capacitive (33 nF) feed back so that it may cut treble. The final amplifier of 3W is similar to the one shown in Figure 2. The quality of reception improves with the size of the loud-speaker. Its diameter should at least be 10 cm. When the circuit shown is used both talking stations hear simultaneously. If it is required to call only one station, then each station must have a separate switch to the loudspeaker line. It is suggested that multi-core cable be used when installing this intercom.

PROPORTIONAL THERMOSTATIC CONTROL

The temperature of water baths or chemical baths may be regulated by thermostats. But during switching a hysteresis appears and the temperature of the bath fluctuates. With proportional control the constancy of temperature may be improved by more than ten times. The circuit in Figure 47 doesn't show the noise killer necessary for half wave control.

REV COUNTER FOR PETROL ENGINE

The Rev counter for petrol engines consists mainly of a one shot stage. As constant temperature is required, it is better to construct the circuit with ICs than with transistors. The interrupter contact gives a pulse which passes the 100 pF capacitor to a monostable stage. The proportion of time per pulse is dependent on frequency. The arriving pulses are added by the meter and give analogue information which is proportional to the speed of the motor. The scale may be calibrated by varying the trimmer 5 kΩ while the speedometer is connected to the square wave generator as is shown in Figure 23. It should be noted that each cylinder of a four stroke motor produces 1 pulse after every second rotation. That means a four cylinder, four stroke motor produces 2 pulses per rotation: 2000 rotations per minute will then produce 100 Hz.

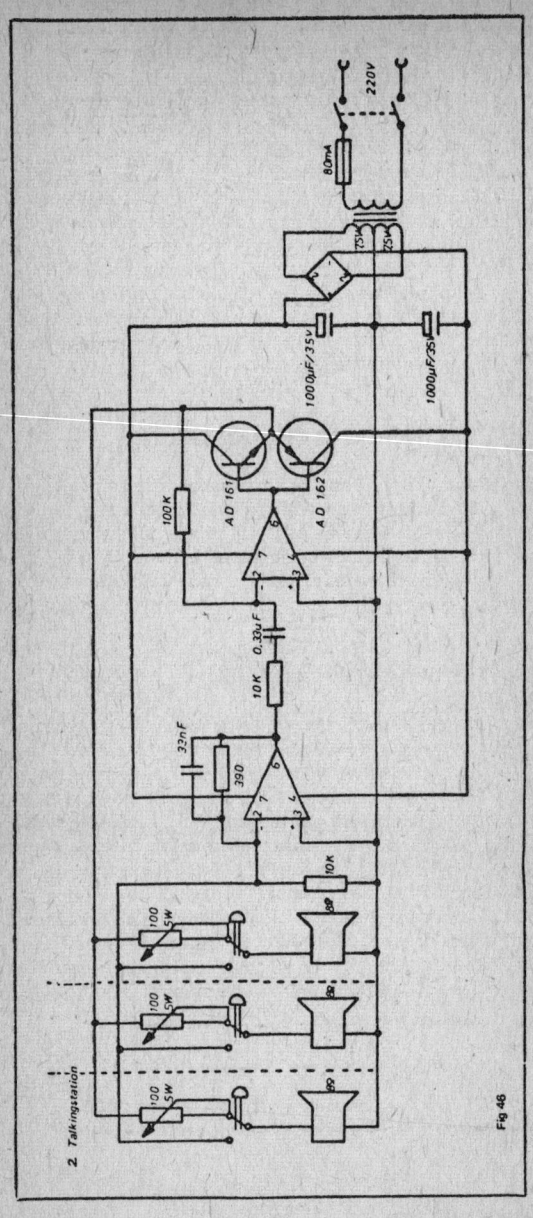

2 Talkingstation

Fig 46

58

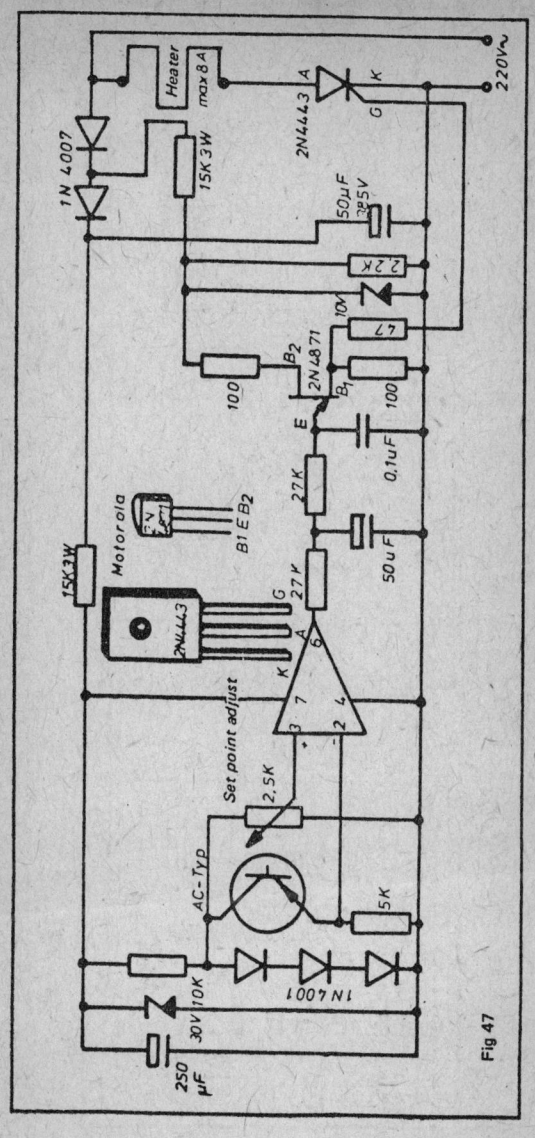

Fig 47

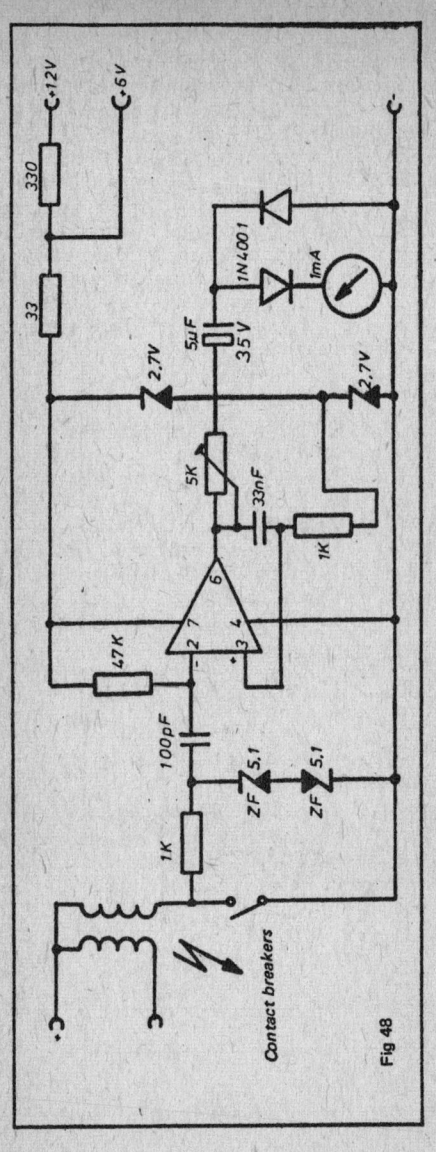

Fig 48

REV COUNTER FOR DIESEL ENGINE

It is also possible to make an electronic rev counter for a Diesel Engine
The Petrol engine has contact breaker points from which the pulses
to the monoflip are obtained. The Diesel engine, being compression
fired, and thus not having spark plugs and contact breaker can not give
the necessary pulses. The circuit shown in Figure 49 shows how the
rev counter may be used on a Diesel Engine. In the case of the phototran-
sistor is built in, which is illuminated by a little 12 V lamp. Pulses
proportional to the rotation are produced when the cooling fan in-
terrupts the light beam. Calibration is done in the same way as
previously mentioned, but it should be noted how many blades the
fan has. as this effects the calibration. Also one must determine
whether the fan is geared up in any way or whether it rotates once for
each rotation of the crankshaft.

TOUCH SENSOR

Switching effects may be initiated by touching the contacts K_1 and K_2.
As the contact resistance of the finger-tip for two contacts in parallel
connection amounts to 50 kΩ, the pin 3 of the IC goes positive while 2
connected to half of supply voltages does not. Output 6 goes positive
also and BC 140 opens. The relay is energized while the finger touches
the two contacts at the same time. The 10-MΩ-resistor from 6 to 3
causes it to function in the way a Schmitt-trigger does. When mains
operation is used, it is sufficient to touch the contact K_2, K_1 not being
necessary.

MASTER CONTROL FOR WARNING FLASH SIGNAL

The law, in certain circumstances, sometimes requires, lights to flash
within limits of frequency e.g. car trafficators. With transistors circuits
it is nearly impossible to meet this demand because temperature drift
occurs for normal working temperatures. As far as temperature conditions
are concerned ICs surpass the transistor circuits and are therefore very
suitable for this purpose. It is not necessary to stabilize the supply
voltage, as time depends only on the 50 μF capacitor and the 50 KΩ
trimmer.

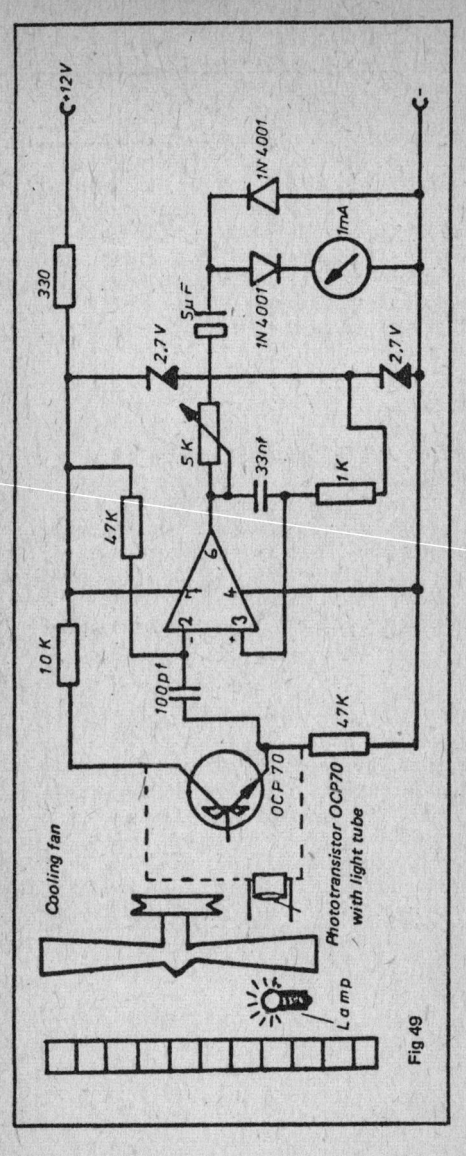

Fig 49

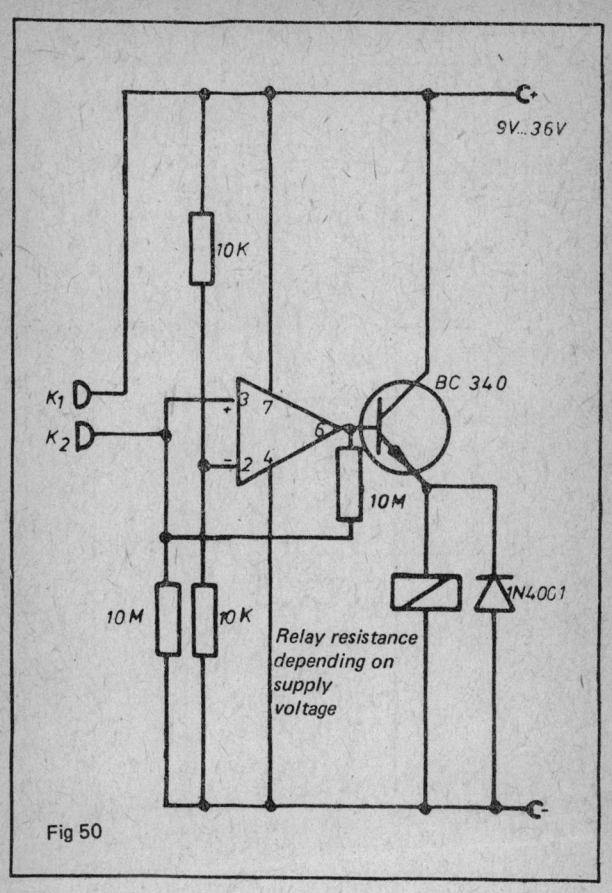

Fig 50

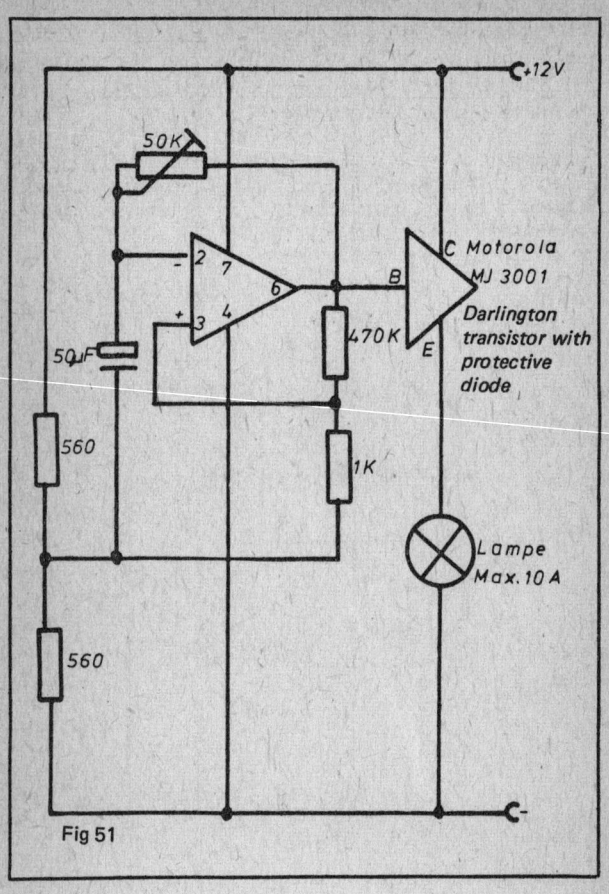

+12V

50K

C Motorola
MJ 3001

Darlington
transistor with
protective
diode

2 7
−
6   B
3 4
+   E

50μF

470K

560

1K

Lampe
Max. 10 A

560

−

Fig 51

WIND SCREEN WIPER AUTOMATIC

Many commercial windscreen wiper timers are available on the market, however it is quite easy for the amateur to construct one himself.

Basically the instrument consists of an non symmetrical multivibrator which operates the high capacity of the timing capacitor.

As the pulse duration is adjustable, it is possible to let the wind screen wiper work for several sweeps without interruption. Certainly this is advantageous in comparison to a circuit which only manages to initiate a single sweep and then stop operating.

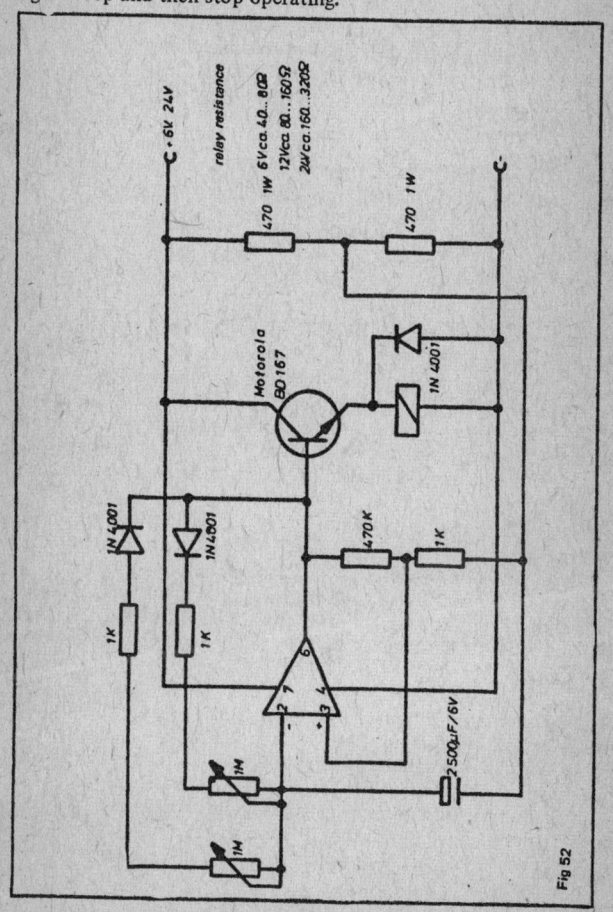

Fig 52

LINEAR OHM METER

To calibrate the non-linear scale on a home made ohmmeter is hardly practical for the home constructor to undertake. This is due, in general, to the lack of high precision standard resistors of the necessary tight tolerance of ½ or 1%, and the high expense of the same, also the time consuming work involved in the adjustment and marking of the dial.

Circuit 53 shows a linear ohmmeter which may be used with any multi-meter or with the circuit of the analogue to digital convertor, shown in Fig. 33 of this book, with a suitable display.

The voltage supply should be stabilised for accuracy. It should be noted that the Trimmer-resistor bank is in the form of a decreasing power decade box, each range being one tenth of the previous.

In the highest range $Rx \leqslant 100M\Omega$.

Initially Rx should be shorted out, for each range, and the potentio-meter P_1 adjusted so that a zero reading is shown on the meter.

STROBOSCOPE

For measurement of unknown machine speeds or the observation of fast processes of machine tools, stroboscopes have been employed very successfully for many years. Recently stroboscopes have been used in some Discotheques and clubs. For this application simple circuits with low frequency constancy are needed, but for technical application an exact frequency is needed, otherwise no standing images are obtained. The multivibrator, designed with an IC, gives a constant frequency which can be adjusted with the $1M\Omega$ potentiometer. To avoid hum interference it is advisable to provide the IC with screening. Any flash tube available on the market which is suitable for stroboscopic applica-tion may be used. Photo flash tubes, often offered, as stroboscope tubes, have a very short lifetime of maybe only 10 minutes to 2 hours and are therefore in the long run no cheaper than the initially more expensive proper stroboscope tubes with a lifetime of about 100 up to 300 hours.

The maximum power of the flash tube should not be exceeded, it is given in Wattsec. Flash lamps from 50 to 400 Wsec. are normally available. Because the circuit operates with voltage duplication directly from mains line and voltages of up to 620VDC are obtained, under no circumstances must the circuit be touched while connected to the mains supply. The case must be suitably insulated and earthed and the shaft of the potentiometer must be fitted with an insulated extension.

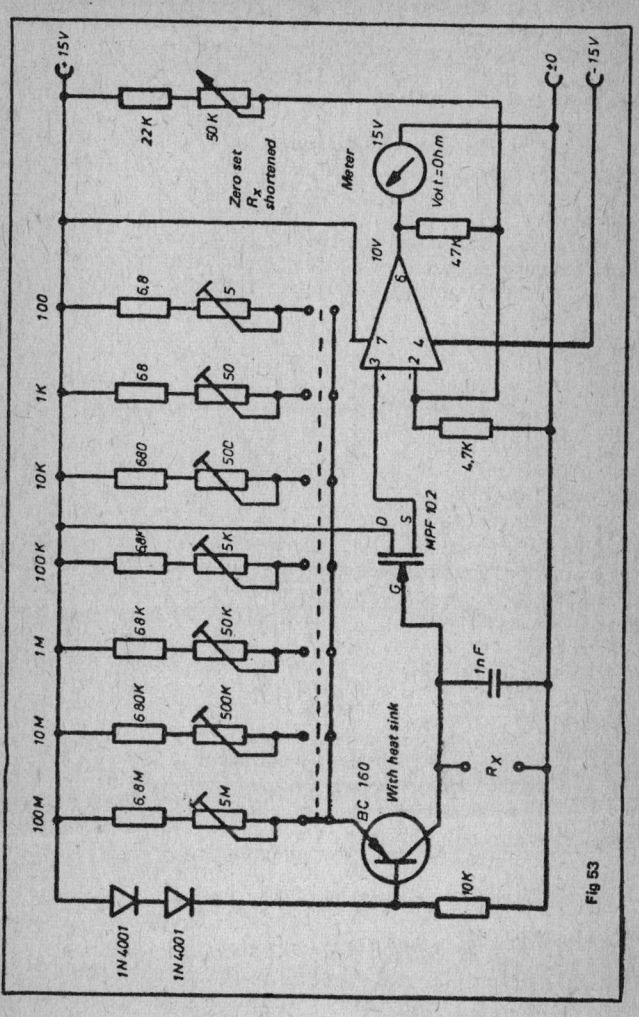

Fig 53

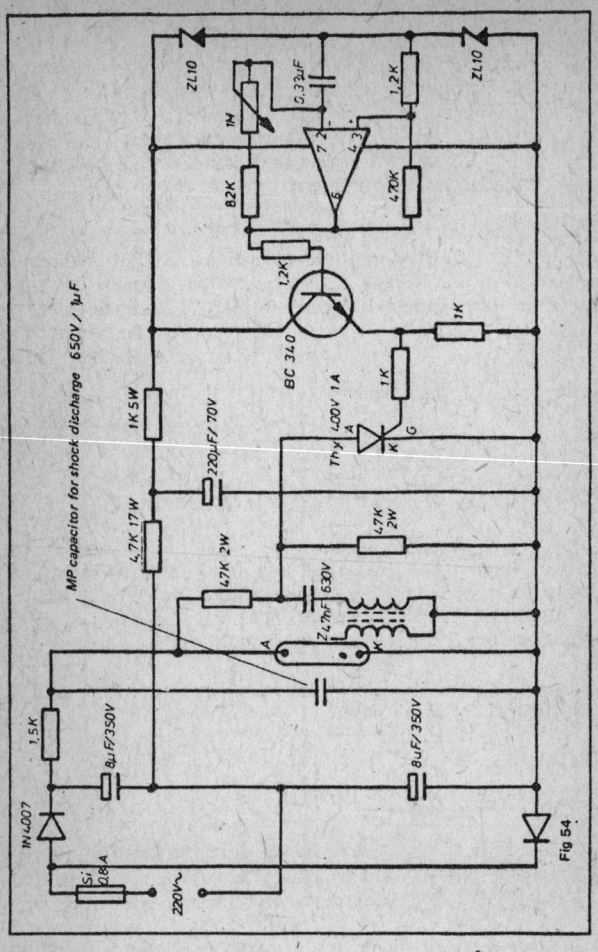

Fig 54

The power can be evaluated with formula $Nw_s = \dfrac{u^2}{2} \cdot C_{\text{farad}} \cdot f$

The following values should be used in Circuit 54
u = 620 volts
C_f = 0.00001 Farad = 10uF
f = flash frequency 25/sec.
Then the power is:

$$\frac{620 \times 620}{2} \times 0.00001 \times 25 \underset{=}{\Omega} 50\ Ws$$

15W HI-FI AMPLIFIER FOR 12V SUPPLY

For sporting events or meetings often amplifiers with higher power must be operated from the batteries of a car or a motorboat. If quadruple power is required the given circuit with the same layout can be supplied with 24V. Modern Darlington transistors are used for this application but they are replaceable by usual types.

Current adjustment which should be about 20mA, is by the $1k\Omega$ trimming resistor. The transistors must be insulated and fitted to a common heat sink to provide sufficient heat conduction. Finger heat sinks about 30 to 40mm high are suitable. The transistor BD167 can be fastened between the fingers of the MJ1000 or MJ900 heat sink. The input sensitivity is $0.6V_{rms}$. The harmonic distortion less than 0.3% for 1kHz.

HI-FI AMPLIFIER OF 12 WATT

A combination of the circuit No. 20 and No. 18 forms the circuit No. 56. The supply current without input signal reaches 10mA and thus only the output transistors produce no heat. The double copperplated circuit board is sufficient as heat sink. However, attention should be paid to the heat contact between the transistor housing and the board.

Supply voltages between 6 and 24V can be used as shown in the Table without any variation of the circuit.

CHANGING LIGHT SWITCH

The light switch given in Fig. 32 is not useable for many purposes because constant illumination by daylight or ambient light prevents its function. Circuit No. 59 shows a chopped light bar with transmitter and receiver.

The transmitter consists of a multivibrator which fires a thyristor 200 times p. sec. and in this way a laser diode is driven. This diode emits strong pensiled rays of infrared light which is invisible to the eye.

The receiver consists of a photo transistor and an ac amplifier with following rectifier.

The incidence of chopped light energizes the relay. The receiver cannot be disturbed by foreign light.

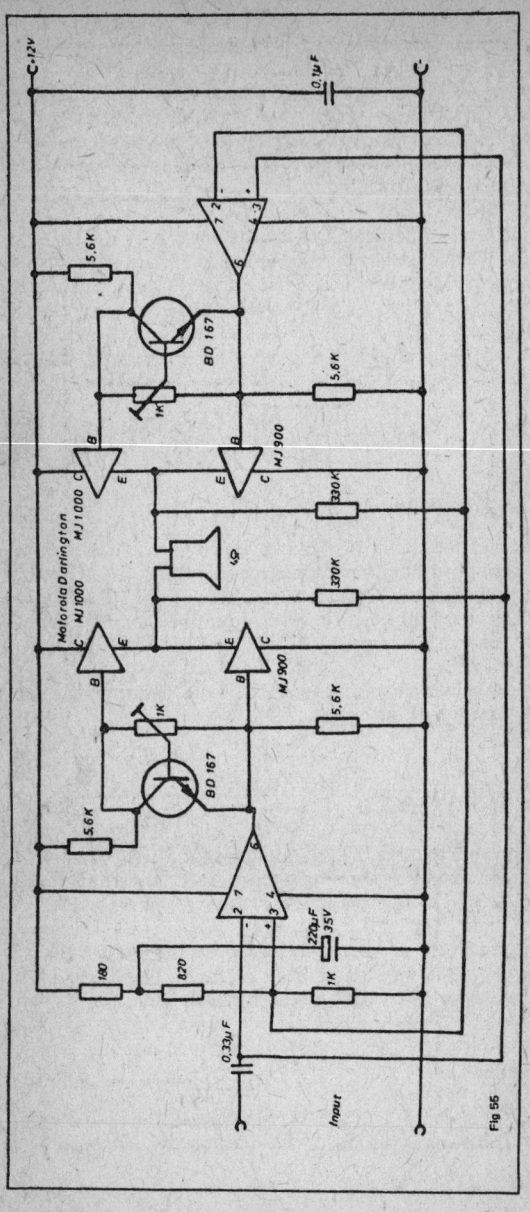

Fig 56

70

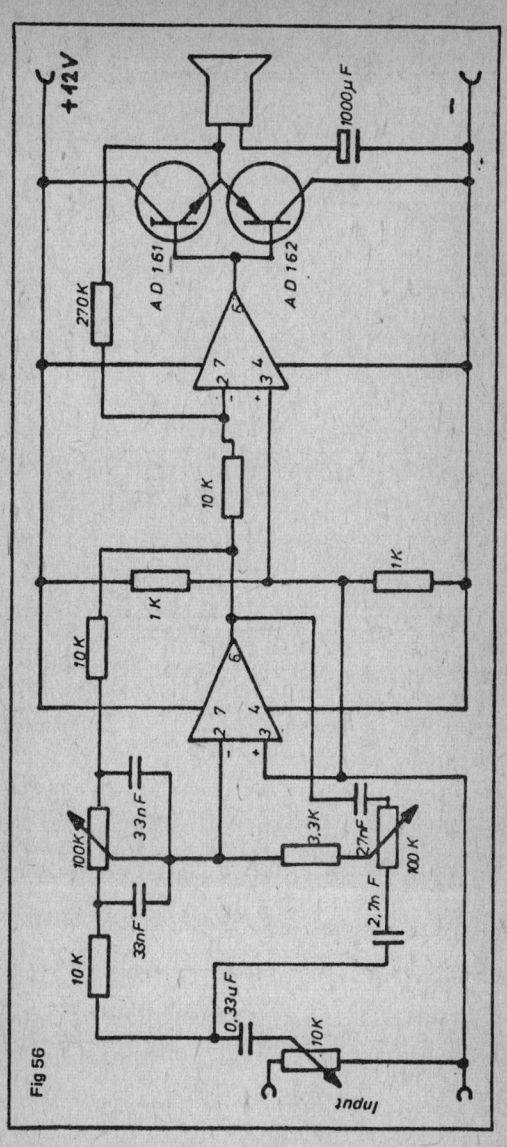

Fig 56

71

Location of the components

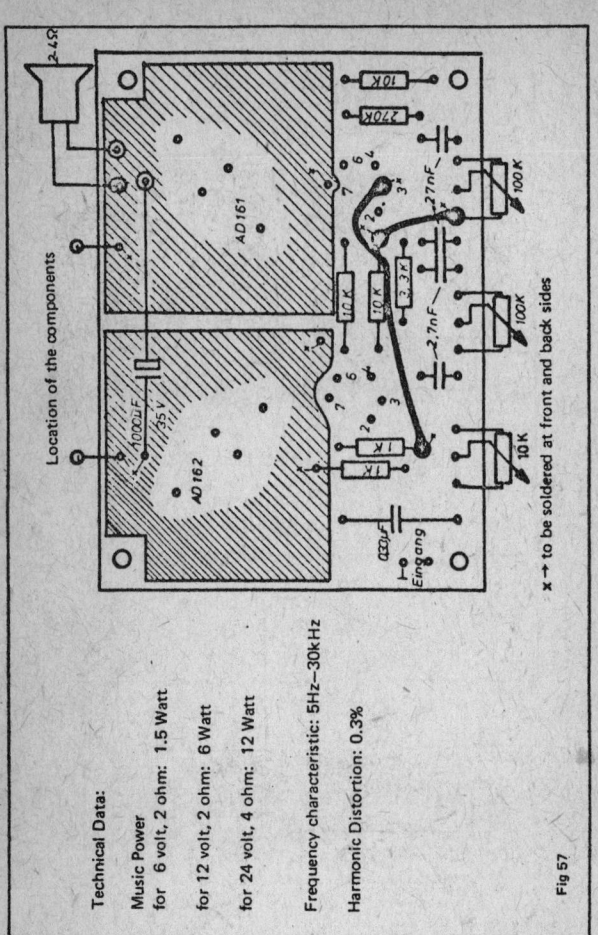

x → to be soldered at front and back sides

Fig 57

Technical Data:

Music Power
for 6 volt, 2 ohm: 1.5 Watt

for 12 volt, 2 ohm: 6 Watt

for 24 volt, 4 ohm: 12 Watt

Frequency characteristic: 5Hz—30kHz

Harmonic Distortion: 0.3%

72

Double copperplate Board for the 15W Hi Fi Amplifier Component Side   Fig 58

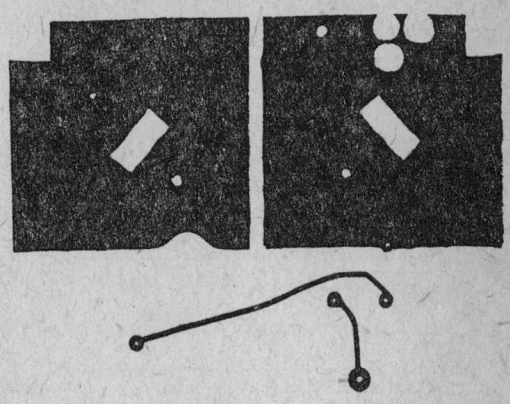

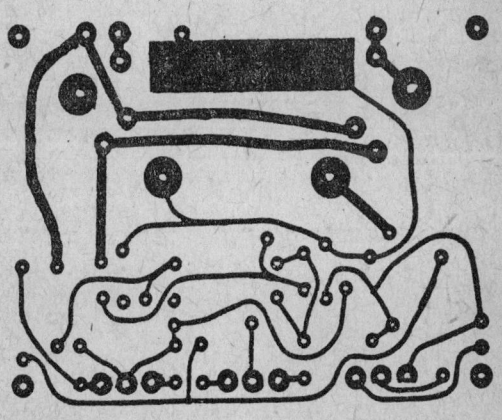

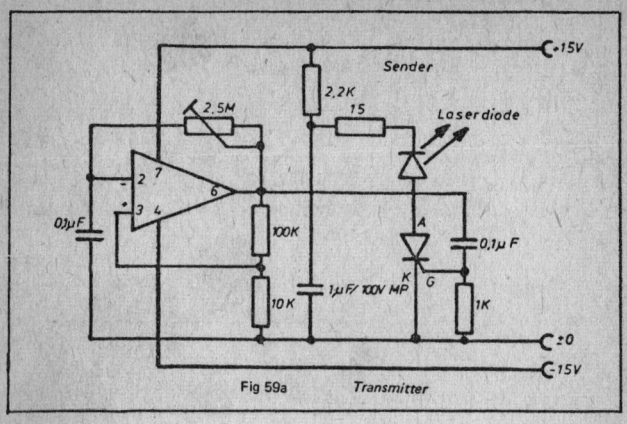

Fig 59a   Transmitter

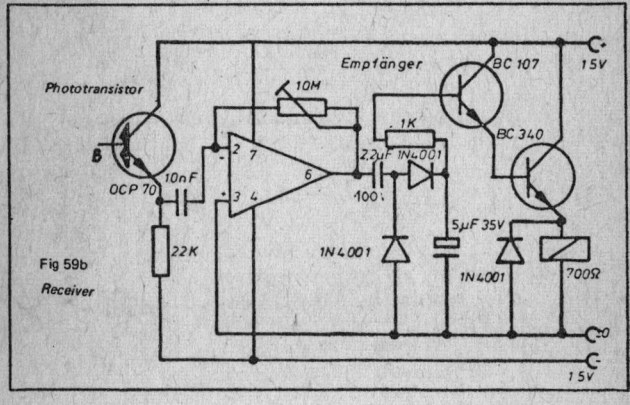

Fig 59b
Receiver

GUITAR PREAMPLIFIER

The preamplifier of Fig. 60 is constructed around an IC 741 and should be connected to the input of a usual guitar amplifier.

The signal voltage comes from the volume voltage divider and is supplied to the non inverting input of the 741. Sound control is by a frequency dependent r.c. network allowing feed back from the output to the inverted input of the IC. The frequency characteristic of the feed back is adjustable, the potentiometer functions as a "taste knob" to determine the "personal sound".

Disturbing parts of the noise can be suppressed by the switch S_1 .

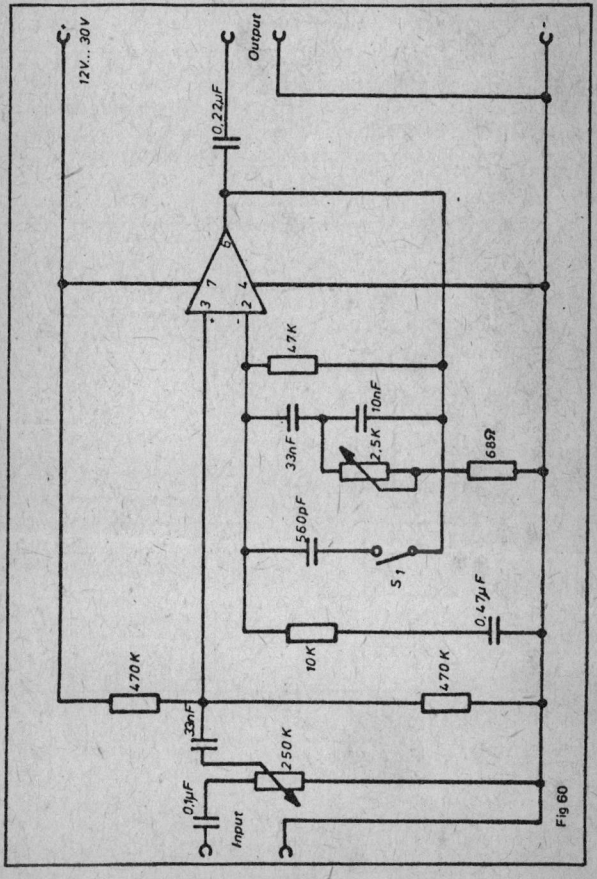

THE VOLTAGE SUPPLY

The described amplifier circuits needs two supply voltages. Because of the small load by the IC, zero voltage can be obtained by resistors or two Z-diodes (Figs. 64/65). The most simple circuit is shown in Fig. 61. For most experiments, this circuit is sufficient, but not for applications where stabilisation is required. The Wien bridge of Fig. 26 for instance can only be operated with a carefully stabilised supply. Audio amplifiers, with feed-back, can suffer from "rumbling" and "motorboat" noises if the preamplifier supply voltage is obtained from the power amplifier supply. Appropriate decoupling by RC networks, with high time constants, should be used as shown in Fig. 66, or better still is to have a separate preamplifier supply, by means of say, an additional winding on the mains transformer.

CIRCUIT 61

advantages: simple construction
               no hum interferences, because battery operated
disadvantages: varying voltage with load and lifetime of the batteries.
               For continuous operation very expensive. Only suitable
               for experiments and portable instruments.

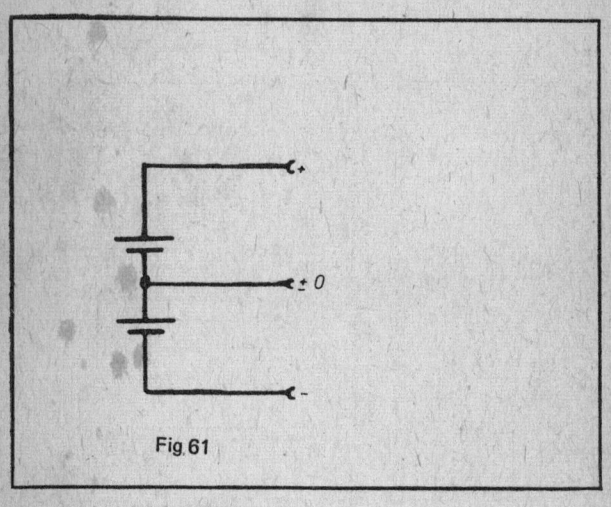

Fig. 61

CIRCUIT 62

advantages: few components. Simple transformer (bell transformer with one winding) required.

disadvantages: high residual line hum, because of half wave rectification on positive and negative voltage. By adding one more capacitor and a smoothing resistor this disadvantage is overcome. The output load may be up to 50mA for the given value of the electrolytic capacitors.

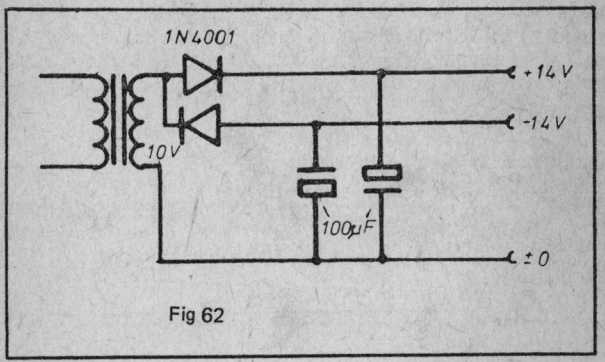

Fig 62

CIRCUIT 63

advantages: full wave rectification demands less capacity of the capacitors for the same residual hum compared to Circuit 62. With more capacity and an appropriate rectifier bridge this circuit can supply up to 10A if the transformer is heavy enough for this load.

disadvantages: transformer with two equal secondary windings is needed.

CIRCUIT 64

advantages: transformer needs only one winding. Only one capacitor required. Voltage constant enough for most needs.

disadvantages: zero voltage line can supply less than 20mA. No more than two ICs can be operated.

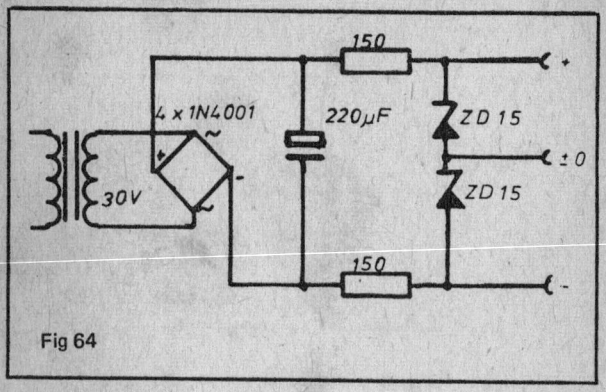

Fig 64

CIRCUIT 65

advantages: compared to No. 64 less expensive.

disadvantages: voltage not stabilised and also disadvantages of No. 64.

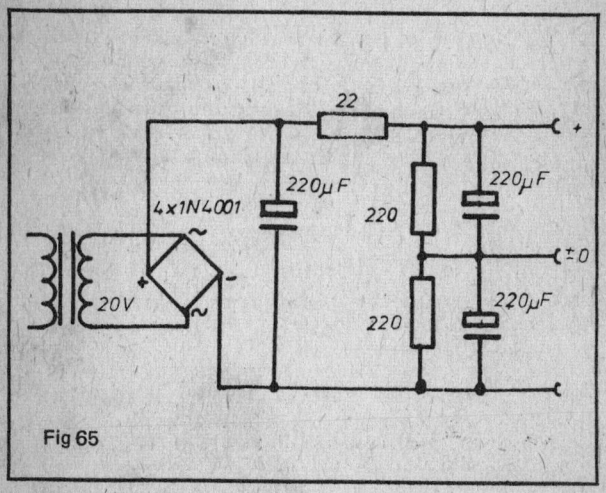

Fig 65

advantages. transformer not necessary if dc voltage is available, e.g. in valve amplifier.

disadvantages: only one or two ICs can be operated.

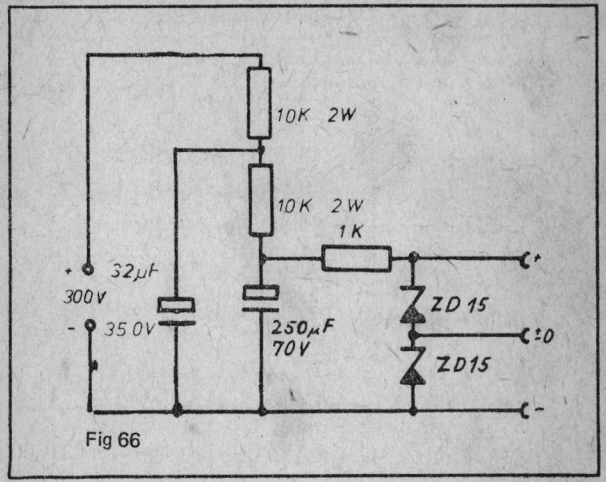

Fig 66

ZERO VOLTAGE CONTROL

Thyristors and triacs are mostly used in an on-off mode in AC circuits.

In the case of phase control switching high radio and TV interference occurs especially with, for example, home constructed electronic light effect units which are rarely properly suppressed. With these, the interference pulses can often affect reception in houses 3 or 4 doors away.

The pulses are difficult to suppress due to the fast rise time and large bandwidth.

With zero voltage control the thyristor or triac is switched on during the zero point of the alternating current. The voltage and current rise proceeds then as a sine waveform.

Power is controlled by on-off switching during a varying number of cycles. For example, two cycles are supplied to the user and the next eight cycles are bypassed, which means that in this example 20% of full power is supplied.

Such good regulation of power can be effected with negative pulses connected to transistor Tr1 . If Tr1 conducts, the IC output cannot go negative during the zero crossing of the AC voltage and the thyristor is not fired.

The on-off pulses can be derived from many different sources for example electronic light effect units, thermostats, switches, etc.

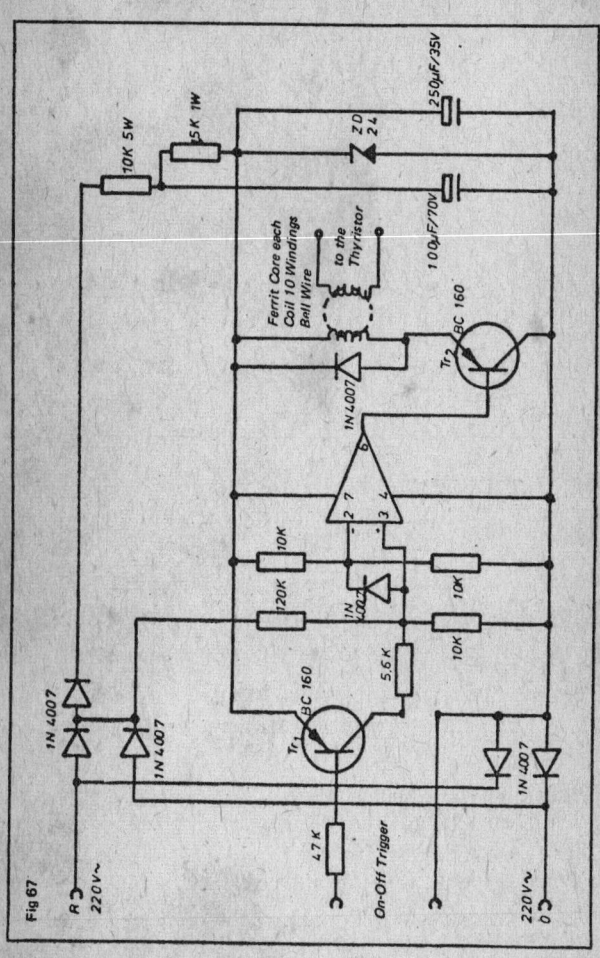

Fig 67